Exemptions and Fair Use in Copyright

EXEMPTIONS
AND FAIR USE
IN COPYRIGHT

The Exclusive Rights Tensions
in the 1976 Copyright Act

LEON E. SELTZER

Harvard University Press

Cambridge, Massachusetts

and London, England

1978

Library of Congress Cataloging in Publication Data
Seltzer, Leon E.
    Exemptions and fair use in copyright.

    "First appeared in the April and June 1977 issues of
the Bulletin of the Copyright Society of the U.S.A."
    Includes index.
    1. Copyright—United States. 2. Fair use (Copy-
right)—United States. 3. Photocopying processes—
Fair use (Copyright)—United States. I. Title.
KF3020.S44   346'.73'0482   77-13676
ISBN 0-674-27335-4

To Lenore

Preface

In December 1975 I began a ten-month leave of absence, with the aid of a Guggenheim Fellowship, to begin a book on the relationship of the copyright scheme to the economics of book publishing. I set aside the first two months to write a short essay on fair use, something that I would ordinarily have dealt with only later in the book but that I was moved to address then because of the accelerated progress of the Copyright Revision Bill through Congress. As so often happens, however, the essay did not work in the narrow terms in which I first attempted it: I had to look more deeply into several interrelated aspects of copyright. Before I knew it my essay ran to book length, my Guggenheim year was over, and the Copyright Revision Bill that I had been struggling to analyze was the Copyright Act of 1976. But the passage from Bill to Act did nothing to resolve the troublesome questions of fair use and exempted use that I had been wrestling with, and it became clear that my treatise—different though it was from the elegant book I had set out to write—might at least have its uses in the continuing public policy debate on the proper limits of copyright protection.

This book first appeared in the April and June 1977 issues of the *Bulletin of the Copyright Society of the U.S.A.*, edited by Alan Latman of the New York University School of Law, to whom I am much indebted for his reading of the work and for his dispatch in arranging the extraordinary measures by which the piece could appear so promptly and completely. Except for the addition of the appendixes, the correction of a few typographical errors, and some minor textual editing, the text here is unchanged.

Marc A. Franklin of the Stanford University School of Law and Deborah S. Griffin of the Massachusetts Bar kindly read an earlier draft of the manuscript and gave me helpful counsel, as did J. G. Bell, editor of Stanford University Press, to whose skilled editorial pencil the first chapter owes much clarity. I am grateful, too, to Charlene Androes, whose usual magic converted my manuscript from nearly indecipherable scribbles into pristine typescript.

I am deeply indebted to the John Simon Guggenheim Memorial Foundation not only for the fellowship that gave me the time to embark on this study but also for its tolerance, as Gordon Ray amiably explained, of serendipitous departures from original schedules of work. When I sought a place to work undisturbed, Gardner Lindzey kindly offered me the remarkable hospitality of the Center for Advanced Study in the Behavioral Sciences at Stanford. I thank him warmly.

As for my greatest debt, to my wife, the dedication, astringent as it is, says it all. She will know the many ways in which I owe this book to her; there is no way to tell anyone else.

Contents

Exemptions and Fair Use in Copyright

Introduction

Though the empowering copyright clause[1] in the Constitution speaks of granting an author the exclusive right to the use of his work, that right is in practice modified by two restrictions. One is such exemptions from copyright control as Congress might expressly make to accommodate competing constitutional interests. The other is an implied right to use copyrighted materials in certain ways without permission, a right that has come to be known as "fair use."[2] The 1976 Copyright Act[3] differs greatly from the Copyright Act of 1909[4] with respect to both restrictions. It includes for the first time an explicit mention of fair use,[5] hitherto entirely a judicial doctrine; and unlike all previous copyright legislation, it is laden with exemptions.[6]

This book examines the possible impact of the 1976 changes on the workings of copyright in the United States. This inquiry for two reasons requires a setting out of what the copyright scheme consists in. First, since the notion of fair use has to do with the accommodation by the courts of tensions *within* the contemplated workings of the copyright scheme, the major premises underlying those tensions must be examined with some care. Second, since an exemption represents a

---

[1] "Congress shall have power. . .to promote the progress of science and useful arts, by securing, for limited times, to authors and inventors, the exclusive right to their respective writings and discoveries." U.S. CONST. art. I, § 8.

[2] *See* Chapter II, *infra.*

[3] Public Law 94-553 (90 Stat. 2451), Copyright Act of 1976, 17 USC. [Hereafter cited as "Copyright Act."]

[4] Act of March 4, 1909 [Pub. L. No. 349], ch. 16, 4 Stat. 436.

[5] Copyright Act, § 107. *See* text at n. 50, *infra.*

[6] Copyright Act, § 107-118.

decision by Congress that in a particular instance the copyright scheme cannot be relied upon to work in the public interest, there must first be a clear idea of the workings of the mechanism to be departed from. Accordingly, we shall in Chapter I examine briefly the major premises of the copyright scheme as it has developed historically. Chapter II will look at how the 1976 act deals with the doctrine of fair use. Chapters III and IV analyze the nature and effect of the new act's unprecedented exemptions from copyright. Chapter V explores the ways in which the 1976 act, or any other copyright legislation, reconciles the public policy tensions set up by the awarding of "exclusive right" to an author in the use of his work.

I.

The Nature of the
Copyright Scheme

A. The Basic Tensions

The copyright scheme requires the accommodation of two distinct sets of opposed principles, one having to do with access, the other having to do with economic costs. With respect to access, the tension is between on the one hand the general principle of the freest possible dissemination of knowledge, and on the other, the copyright-scheme restrictions on access to works of the intellect. With respect to cost, the tension is between on the one hand the general principle of the maximum freedom of competition, and on the other the kind of monopoly-like economic restrictions intrinsic to the copyright design.

The success of a legislature in fashioning a copyright statute will necessarily depend on how clearly the interplay of these tensions within the copyright scheme is understood. The tensions are the central focus of this book. But before we undertake to examine them, we must dispose of a preliminary question, namely, whether copyright is in fact a legitimate instrument of public policy. The question can be approached in either of two ways. One, taking its orientation from traditional notions of private property, sees copyright either as protecting existing "natural rights" of the author or as properly establishing authors' rights where they had not existed before.[7] The other

---

[7] The classic argument took place in the cases of Millar v. Taylor, 4 Burr. 2303, 98 Eng. Rep. 201, and Donaldson v. Beckett, 4 Burr. 2408, 98 Eng. Rep. 257; and in the United States in Wheaton v. Peters, 33 U.S. (8 Pet.) 591 (1834) [only the Brightly (1884) third edition of Peters carries the text of the dissents], where the Supreme Court held that the statute created the author's rights. Both *Millar* and *Donaldson* held that the right existed at

views copyright as an unwarranted monopoly, and sees its origin in such suspect matters as religious censorship, royal printing patents, state control of political dissent,[8] and the protection of special interests like those of artisans in certain guilds or those of booksellers in cartel-like associations, reason enough to find copyright inherently suspect.[9]

Proponents of this second view question on neo-classical economic grounds[10] the major public-policy premise of copyright—namely, that it is the most efficient way to provide society with the tangible fruits of intellectual and creative work—and support their opinion with one or more of three arguments. First, they maintain that the monopoly-like features of copyright, by restricting competition, exact an exorbitant "monopoly subsidy" in the form of an exorbitantly high price for copyrighted materials, a price unjustified by the offsetting advantage, if any, to society. Second, they argue that the publisher's intermediary role between author and society so alters the economic facts as to render untenable the copyright scheme's reliance on author incentive. Third, they claim that new modes of disseminating copyrighted materials—sound recording, television, photocopying, computer storage and retrieval—have so changed the dimensions and configuration of the public need for access to published materials as to render the copyright scheme obsolete.[11]

Each of these arguments urges the ultimate dominance, over the author's "exclusive right," of one or both of the basic principles of public policy—freedom of competition and freedom of information—with which the copyright scheme tries to reconcile that right. In effect, they argue that when in order to further the creation and dissemination of works of literature, knowledge, and the arts, an author is given an economic incentive in the form of "exclusive right"

---

common law, but *Donaldson* went on to decide that once a copyright statute was enacted it superseded the common-law rights. The straightforward "private property" theory underlain by the "natural right" of authors to the fruit of their labors is set out in *Wheaton* and rejected by Justice Story as the governing rationale for the constitutional grant. "Nevertheless," says Professor Nimmer, "there is nothing to indicate that the Framers in recognizing copyright intended any higher standard of creation in terms of serving the public interest than that required for other forms of personal property." NIMMER ON COPYRIGHT § 3.1.

[8] "[C]opyright has the look of being gradually secreted in the interstices of the censorship." B. KAPLAN, AN UNHURRIED VIEW OF COPYRIGHT 4 (1967).

[9] *See* A. Plant, *The Economic Aspects of Copyright in Books,* I ECONOMICA 167 (new series 1934) [hereafter cited as "Plant"].

[10] *See* notes 15 to 18, *infra,* and accompanying text.

[11] *See e.g.,* Henry, *Copyright: Its Adequacy in Technological Societies,* 186 SCIENCE 993 (Dec. 13, 1974).

to his work, the price to the public, in terms of cost or access or both, must be excessive. In short, they argue, copyright is simply a monopoly like any other monopoly, and its commodity—the book, say, or the specialized journal—is governed in the marketplace by the classical relationship between price, supply, and demand. It is accordingly in the public interest for Adam Smith's "invisible hand"—in which market forces produce both the goods themselves and the lowest prices for them—to operate without copyright constraints.

This argument goes back at least as far as the booksellers' petition to the Long Parliament in 1643.[12] It was prominent in the debate about the licensing Act of 1662,[13] which led to the first copyright law, the Statute of Anne,[14] in 1709. It has been made more recently in the copyright hearings and publications attendant on the inquiry of the British Royal Copyright Commission in 1876-78,[15] in an influential essay by the distinguished economist Arnold Plant in 1934,[16] in an article by the American economists Hurt and Schuchman in 1965,[17] and notably in a 1970 article by Professor Stephen Breyer of Harvard.[18]

The validity of such arguments, resting as it must on a careful examination of the relevant economic facts and of the applicability of conventional microeconomic models to the behavior of books and journals—and other forms of expression—in the marketplace, is not within the scope of this book. We are here simply concerned with how the 1976 Act supports or undermines the validity of the copyright scheme. We do not deal at all with the first two arguments—"monopoly" effects and the entrepreneur's role. The third argument, that the copyright scheme has been rendered invalid by technological change, we deal with only insofar as technological change figures explicitly in the 1976 Act.

---

[12] The petition appears in I ARBER, A TRANSCRIPT OF THE REGISTERS OF THE COMPANY OF STATIONERS OF LONDON 584 (1875). *See* BLAGDEN, THE STATIONERS' COMPANY: A HISTORY 1403-1959 (1960); L. R. PATTERSON, COPYRIGHT IN HISTORICAL PERSPECTIVE 128-31 (1968).

[13] 13 and 14 Car. II, c. 33.

[14] 8 Anne, c. 19 (1709).

[15] Sir Louis Mallet, in Royal Commission on Copyright, The Royal Commissions and the Report of the Commissioners, p. xlvii (London 1878); T. H. Farrer, *Fortnightly Review* 836 (December 1878); and a review on the Report of the Copyright Commission appearing in the *Edinburgh Review* 295 (October, 1878).

[16] *Supra*, n. 8.

[17] Hurt and Schuchman, *The Economic Rationale of Copyright*, 56 AM. ECON. REV., May, 1966 (1965 Papers and Proceedings of the Amer. Econ. Ass'n).

[18] Breyer, *The Uneasy Case for Copyright*, 84 HARV. L. R. 281 (1970) [hereafter cited as "Breyer"]

But at least four reasons argue that it is not too rash to accept the copyright scheme as valid for our purposes.

First, there is simply no evidence to support the theoretical argument, which is entirely speculative. On the one hand, we see in the United States (and other copyright countries) what appears to be a wholly uninhibited proliferation of books and journals, with a wide range of prices. So far as books are concerned, there are substantial numbers at both ends of the price spectrum, with the most visible books—paperbacks—the most widely accessible and the cheapest.[19] At the very least, copyright cannot easily be seen to have inhibited the production either of new works of the intellect or of inexpensive books. On the other hand, there is no sophisticated society where books and journals are freely produced in uninhibited abundance that does not have a copyright scheme. There are accordingly no examples of the supposed economic benefits of a noncopyright system.

Second, the number of conventional microeconomic rules that do not appear to fit the marketplace behavior of publishing raises serious questions about either the accuracy of the usual factual analysis of publishing economics or the adequacy of the economic models imposed on the facts, or both.[20] The consequence of this failure of a

---

[19] There are in the United States some 429,000 books in print, with some 40,000 new books being published annually. BOWKER, BOOKS IN PRINT (1975), BOWKER, ANNUAL LIBRARY AND BOOK TRADE INFORMATION (1975). The Harvard University Library subscribes to 100,000 periodicals. Rutherford D. Rogers, Librarian of Yale University, quoted in PUBLISHERS WEEKLY, August 9, 1976. Paperback titles in print in 1976 numbered some 132,000, with some 13,000 new paperbacks being issued annually. BOWKER, PAPERBOUND BOOKS IN PRINT (1976), BOWKER, ANNUAL LIBRARY AND BOOK TRADE INFORMATION (1975).

[20] The principal reason for what appears to be the intractability of publishing to the models of microeconomic theory is probably that what is characterized as monopoly is in fact vigorous competition: a new novel competes for attention with every other new novel as well as with old novels, with other new books that are not novels as well as with other old books that are not novels, and with an enormous spectrum of other media making claims on the attention of a particular mind. In consequence the usual configurations of price and cost curves do not seem to apply to factors of supply and demand in publishing. For example: that small changes in price would result in increases in demand, and therefore in supply, does not appear offhand to apply to books; the exclusive controller of supply, rather than optimizing his marginal revenue by keeping prices high, as conventional theory would have him do, instead often puts out his product in cheaper and cheaper editions—first in book-club editions, then in paperback; and the usual time factors underlying the standard premises about the effect of either "perfect" or imperfect competition do not seem to apply where suppliers can respond to a market signal by producing ten thousand or ten million units in a matter of days.

theoretical noncopyright system to support analytically the expected conclusion that all of the desirable social ends would thus be obtained, usually requires the invention of alternative schemes to copyright. For example, one such analysis suggests three alternatives to the copyright-scheme incentive and economic mechanisms: government prizes and subsidies instead of author royalties; pre-publication contracts between producers and customers in place of risk-taking for certain kinds of books; and such reward to the author and publisher as may accrue from "lead time," i.e., from the initial sales of a book in the short period before other publishers can issue competing editions of it.[21] When the weakness of these alternatives was pointed out,[22] the essential question was defined as being not whether there should be a copyright scheme at all, but "*how much* protection is appropriate."[23] Since that question is precisely what the 1976 Act or any other copyright act seeks to answer, it would appear that even its critics in some fundamental sense accept the validity of the copyright scheme.

Third, even the educators and librarians, who opposed the "exclusive right" argument most passionately in the long debate in Congress on the Copyright Revision Bill, rarely questioned the fundamental validity of copyright as such. Rather, they argued that the superior claims of education and of library service for free access could be honored without damage to the scheme.[24]

And finally, Congress has simply chosen not to abandon the copyright scheme.

Accordingly, then, we may without apology regard the existing copyright scheme as having persuasive claims to validity. It follows that for our limited purposes we do not care whether the author's property-like interests in his work are "natural" or given him by the state, we do not care about the origins of the state's interest in copyright-like controls, and we do not care whether or not the whole copyright notion is defensible in neoclassical economic terms. The Constitution and the statute define the author's interests, express the state's purpose, and establish the economic framework of the copyright scheme; that is enough. We turn next to the Constitutional basis of copyright.

---

[21] Breyer, *supra* n. 18.

[22] Tyerman, *The Economic Rationale for Copyright Protection for Published Books: A Reply to Professor Breyer*, 18 UCLA L. REV. 1100, 19 BULL. CR. SOC. 99, Item 57 (1971).

[23] S. Breyer, *Copyright: A Rejoinder*, 20 UCLA L. REV. 75, 83 (1972). [Emphasis supplied.]

[24] *See* the summary in Register of Copyrights, Second Supplementary Report on the General Revision of the U.S. Copyright Law: 1975 Revision Bill (October-December, 1975), chs II and III.

B. The Meaning of the Constitutional Copyright Clause: The Primary Balancing

The tensions among the contradictions inherent in copyright are initially balanced in Article I, section 8, clause 8, of the Constitution, which gives Congress the power "To promote the progress of science[25] . . . by securing for limited times to authors . . . the exclusive right to their . . . writings." It is thus stated or implied

—that the products of the intellect are to be especially encouraged;
—that the way to do this is to give authors a monetary incentive;
—that there is something about printed materials that prevents our relying on the ordinary workings of the marketplace to ensure their production and distribution at appropriate levels;
—that the economic incentive for the author shall consist in a grant of the exclusive right to make copies of his work;
—that such controls smack of monopoly;
—that a monopoly is inherently against the public interest, because it either denies free access to the work or adds to its cost, or both;
—that on this account it is to be limited in time; and
—that this limitation is sufficient to avoid imposing excessive costs, either of price or of access, on the public.

The central elements of the clause may be stated as follows:
1. The *purpose* of copyright is to benefit society.
2. The *mechanism* by which this purpose is achieved is to be economic.
3. Society's *instrument* in achieving this purpose is to be the author.[26]
We can construct alternative relationships among these elements, but only by ignoring the emphasis embodied in the constitutional text, notably in the phrases "to promote" and "by securing." Despite occasional challenges,[27] this particular arrangement has constituted the theoretical basis for copyright for at least two centuries. Since it is by no means as simple as it sounds, and since a clear understanding of its

---

[25] The term "science" is used in its older meaning as knowledge of all kinds.
[26] The fact that the author's agent, in this conceptualization, is a book or journal publisher, a theatrical producer, a motion-picture company—with each of whom the author makes a contract with respect to his economic interests in his copyright—does not change the fundamental relationship between author and society.
[27] *E.g.*, T. MACAULAY, SPEECHES ON COPYRIGHT (1841 and 1842) (Gaston ed. 1914); A. BURRELL, SEVEN LECTURES IN THE LAW AND HISTORY OF COPYRIGHT IN BOOKS (1899); Plant, *supra* n. 9.

workings is essential if we are to deal with its accommodation of the notions of fair use and exempted uses, let us briefly consider its history as it pertains to the United States.

The first express statement of the present copyright scheme appears in the Statute of Anne, which however complicated the history of its gestation, and however diverse the competing factors in its establishment of public policy, nevertheless is explicit about copyright's purpose and means: "An Act for the encouraging of learning, by vesting the copies of printed books in the authors or purchasers of such copies, during the times therein mentioned."[28] That this was the model for the United States approach is clear from the parallel wording of the purpose clause of the first federal copyright act passed by Congress in 1791: "An Act for the encouraging of learning, by securing the copies of maps, charts, and books to the authors and proprietors of such copies, during the times mentioned therein."[29]

The "purpose" Preamble disappeared from the United States statute with the revision of 1831[30] and has not appeared since. Both the 1909 act and the 1976 act simply begin by defining the elements of the exclusive right conferred on authors by the Constitution. It has nonetheless been necessary for legislatures and courts[31] to confront the purpose question constantly in fashioning copyright legislation. Thus, though the new statute says nothing about the general purpose of copyright either in the text or in the accompanying reports,[32] the legislative committee report on the 1909 bill did. It set out the purpose of copyright as follows:

---

[28] 8 Anne, c. 19.

[29] Act of May 31, 1790, ch. 15, 1 Stat. 124.

[30] Act of Feb. 3, 1831, ch. 16, 4 Stat. 436.

[31] E.g., Mazer v. Stein, 347 U.S. 201, 219 (1954): "The economic philosophy behind the clause empowering Congress to grant patents and copyrights is the conviction that encouragement of individual effort by personal gain is the best way to advance public welfare through the talents of authors and inventors. . ."

[32] House Committee on the Judiciary, House Report No. 94-1476, Sept. 3, 1976, to accompany S. 22, 94th Cong., 2d Sess., and Senate Committee on the Judiciary, Senate Report No. 94-473 to accompany S. 22, 94th Cong., 1st Sess., November 20, 1975, Conference Report, House No. 94-1733, to accompany S. 22, 94th Cong., 2d Sess., Sept. 29, 1976 [hereinafter cited as "House Committee Report", "Senate Committee Report", and "Conference Report", respectively]. *See* Appendixes B and C, *infra*. The introductory section of the committee report on the Copyright Revision Bill passed by the House of Representatives in 1967, however, did have a paragraph on purpose, though it was abandoned in later versions of the report. H. Report No. 83, to accompany H.R. 2512, 90th Cong., 1st Sess., March 8, 1967, p. 3.

"The enactment of copyright legislation by Congress under the terms of the Constitution is not based on any natural right that the author has in his writings, for the Supreme Court has held that such rights as he has are purely statutory rights, but upon the ground that the welfare of the public will be served and progress of science and useful arts will be promoted.... Not primarily for the benefit of the author, but primarily for the benefit of the public, such rights are given. Not that any particular class of citizens, however worthy, may benefit, but because the policy is believed to be for the benefit of the great body of people, in that it will stimulate writing and invention to give some bonus to authors and inventors.

"In enacting a copyright law Congress must consider ... two questions: First, how much will the legislation stimulate the producer and so benefit the public, and second, how much will the monopoly granted be detrimental to the public? The granting of such exclusive rights, under the proper terms and conditions, confers a benefit upon the public that outweighs the evils of the temporary monopoly."[33]

This formulation is essentially an elaboration of the constitutional clause; in particular it accepts the author as the proper instrument of the public purpose. When a court or legislature, however, addresses a copyright question, it is always precisely this element of the scheme that is in question. We must ask, then, just how much the author's interests (which have to do with incentives) do in fact coincide with the interests of the public. James Madison, who alone among the framers of the Constitution has anything to say about copyright, and that very little, deals with it concisely in *The Federalist*: "The public good *fully* ... coincides with the claims of individuals." [Emphasis added.][34] That view, which seems to have been incorporated in both the constitutional clause and the first copyright statute, represents not

---

[33] H. Report No. 2222, 60th Cong., 2d Sess. Quoted also in Report of the Register of Copyrights, Copyright Law Revision, House Committee on the Judiciary, 87th Cong., 1st Sess, p. 5 (Comm. Print 1961).

[34] The full paragraph reads: "The utility of this power will scarcely be questioned. The copyright of authors has been solemnly adjudged, in Great Britain, to be a right of common law. The right to useful inventions seems with equal reason to belong to the inventors. The public good fully coincides in both cases with the claims of individuals. The States cannot separately make effectual provision for either of the cases, and most of them have anticipated the decision of this point, by laws passed at the instance of Congress." The Federalist, No. 43.

only a considerable reliance on the internal economic workings of the copyright scheme but a departure from a more cautious position. The Statute of Anne, for example, hedged its grant of exclusive rights by providing that anyone thinking a book too high priced could bring an action before a tribunal with the power to lower the price and to fine the bookseller.[35]

There is no doubt that the departure was deliberate. On the urging of the Continental Congress,[36] at the suggestion of Madison, the separate states passed state copyright statutes between 1783 and 1786.[37] Five of these statutes,[38] clearly tracking the Statute of Anne, contained provisions for state controls of access to copyrighted works, of their prices, or both.[39] By omitting such provisions, both the Constitution and subsequent congressional copyright statutes accept Madison's argument that the "exclusive right" granted by copyright will not

---

[35] 8 Anne, c. 19.

[36] In a resolution "recommending the several states to secure to the authors or publishers of new books the copyright of such books." 24 JOURNALS, CONTINENTAL CONGRESS 326 (1783).

[37] *See* Copyright Office Bulletin No. 3, U.S. Copyright Office, Copyright Laws of the United States of America (1973).

[38] Those of Connecticut, Georgia, New York, North Carolina, and South Carolina.

[39] The price-control section of the New York statute reads as follows:

"*And whereas* it is equally necessary for the encouragement of learning, that the inhabitants of this State be furnished with useful books at reasonable prices:

"*III. Be it further enacted by the authority aforesaid,* That whenever any such author or proprietor of such book or pamphlet shall neglect to furnish the public with sufficient editions thereof, or shall sell the same at a price unreasonable, and beyond what may be adjudged a sufficient compensation for his or her labour, time, expenses, and risque of sale, any one of the judges of the supreme court of judicature of this State, on complaint made thereof to him in writing, is hereby authorized and impowered to summon such author or proprietor to appear at the next supreme court of judicature, and the said court are hereby authorized and impowered to enquire into the justice of the said complaint, and if the same be found true, to take sufficient recognizance and security of such author or proprietor, conditioned that he or she shall, within such reasonable time as the court shall direct, publish and offer for sale in this State, a sufficient number of copies of such book or pamphlet, at such reasonable price as the said court shall on due consideration affix, and if such author or proprietor shall neglect or refuse to give such security as aforesaid, the said court are hereby authorized and empowered to give such complainant a full and ample licence to re-print and publish such book or pamphlet in such numbers and for such term as the said court shall judge just and reasonable: *Provided,* Such complainant shall give sufficient security before the said court to afford such re-printed edition at such reasonable price as the said court shall thereto affix." New York, Act of April 29, 1786.

impose unacceptable costs to society in terms either of limiting access to published works or of pricing them too high.

C. The Traditional Statutory Design

That early congressional decision to rely wholly on the internal mechanics of the copyright scheme continued to govern the legislative design until 1976. The Copyright Act of 1909, for example, contained no exemptions at all from copyright protection of printed works, only one qualified exemption for making a tangible copy of anything else (a compulsory license, with statutory fees, for making phonograph records of copyrighted music),[40] and two exemptions for certain "performances" of copyrighted work, one for the nonprofit performance of a musical or nondramatic literary work[41] and another for the playing of records on coin-operated machines.[42] All other exceptions to copyright controls were governed by the judicial doctrine of fair use, which we shall shortly examine.[43]

But the coherence of any approach to the question of exceptions to copyright controls, whether as fair use or as statutory exceptions, depends on absolute clarity about the role of the author as the *instrument* in furthering the purposes of the scheme. The common error of classifying the author's interest or society's as "primary" or "secondary," thereby characterizing them as somehow opposed, has often confused analysis, particularly when what is at issue is the conceptualization of the scheme with respect to cause and effect. If the copyright scheme *itself* is to be considered in the public interest, such categorizations blur the fundamental issues usually in question.

This may seem too subtle a point to dwell upon or, perhaps, a distinction not important enough to matter. But we know that the conceptualization of the copyright scheme has always given courts trouble with problems at the margin, such as in fair use and infringement adjudications, and the long and difficult sixteen-year legislative struggle with the questions of fair use and exemptions in the new copyright act reflects a similar difficulty there. This argues for holding on as long as possible, in our analysis, to such distinctions as are to be found in the express wording of the framework and in the history and workings of the scheme.

Consider, for example, the effect of such a perspective on an

---

[40] Act of March 4, 1909, c. 320, 35 Stat. 1075, § 1(e).
[41] *Id.* at § 104; § 1(c) exempts nonprofit performances for "religious purposes," but it is subsumed under the broader scope of § 104.
[42] *Id.* at § 1(e).
[43] *Infra,* Chapter II.

analysis of the alterations in any copyright statute. Insofar as new exceptions simply accommodate the role of the author-as-instrument to the changes in technology, the essential reliance on the workings of the copyright scheme is increased, and therefore insofar as the mechanism is itself valid its efficiency is increased. On the other hand, insofar as the exceptions dilute the controlling role of the author, the reliance on the internal workings of the copyright scheme itself is lessened, and other mechanisms might arguably be seen to be looked to for solutions. Accordingly, we first take a brief look at how the concept of the author-as-instrument has heretofore fared.

We might, for example, find the elements of policy confusion already adumbrated in Lord Mansfield's opinion in *Sayre v. Moore*,[44] where the question was posed in the context of a second writer's use of copyrighted material:

> "We must take care to guard against the two extremes equally prejudicial; the one that men of ability, who have employed their time for the service of the community may not be deprived of their just merits and reward for their ingenuity and labor; the other that the world may not be deprived of improvements nor the progress of the arts retarded."

Lord Mansfield's eloquent and concise statement, unexceptionable as it appears, nevertheless contains those small erosions of relationships that make for an uncertain compass in the development of public policy. Purpose and instrument are no longer quite as distinct as they were. The notion of "progress of the arts," with the seeds of confused purpose innocent but pregnant in the location of the phrase at the *end* of the sentence, comes along after the author's interests are invoked. Is the first author rewarded because he has employed his time "for the service of the community" or because he should "not be deprived of his just merits and reward for his ingenuity and labor?" Perhaps this niggling question ought not to matter, and in most contexts it would not. But if we are looking for signs of the conceptual trouble we know is to come with respect to fair use adjudication, and with legislative exemptions from copyright, they are perhaps here.

It was relatively easy, for example, by 1941 for Professor Chafee to say in an influential article that "the primary purpose of copyright is, of course, to benefit the author,"[45] and for the Supreme Court in

---

[44] 1 East 361 n., 102 Eng. Rep. 139 n. 16-18 (K.B. 1785).
[45] Chafee, *Reflections on the Law of Copyright: I*, 45 COLUM. L. REV. 503, 506 (1945).

1954 to say the opposite in *Mazer v. Stein:* "The copyright law, like the patent statute, makes reward to the owner a secondary consideration."[46] But even the Supreme Court's formulation, though more accurately reflecting the perspective of the constitutional clause, nevertheless introduces the kind of conceptual flaw that is at the root of much of the difficult public debate on the role of the copyright scheme: to say that the benefit to the author is a "secondary consideration" is like saying that when reliance is put on a flask to transport wine across a carpeted room, whether or not the flask leaks is, with respect to getting the wine there, a "secondary consideration."

It is the Supreme Court's imprecise formulation that the Register of Copyrights uses in his 1961 Report orienting Congress on the issues that the revision of the copyright act poses for them:

> "Although the primary purpose of the copyright law is to foster creation and dissemination of intellectual works for the public welfare, it also has an important secondary purpose: To give authors the reward due them for their contribution to society."[47]

The erosion of the concept of the fundamental copyright design that begins from this lack of precision is at once demonstrated by what shortly follows in the Register's report:

> "Within reasonable limits the interests of authors coincide with those of the public. Both will usually benefit from the widest possible dissemination of the author's works. But it is often cumbersome for would-be users to seek out the copyright owner and get his permission. There are many situations in which copyright restrictions would inhibit dissemination, with little or no benefit to the author. And the interest of authors must yield to the public welfare where they conflict.
> "Accordingly, the U.S. copyright law has imposed certain limitations and conditions on copyright protection:
> "—The rights of the copyright owner do not extend to certain uses of the work. . .
> "—The term of copyright is limited, as required by the Constitution. . .

---

[46] 347 U.S. 201, 219 (1954).

[47] Register of Copyrights, 87th Cong., 1st Sess., Copyright Law Revision, Report on the General Revision of the U.S. Copyright Law 5 (House Judiciary Comm. Print 1961).

"—A notice of copyright in published works has been required. . .

"The large mass of published material for which the authors do not wish copyright is thus left free of restrictions.

"—The registration of copyrights and the recordation of transfers of ownership have been required . . . The public is thus given the means of determining the status and ownership of copyright claims.

". . . The ultimate task of the copyright law is to strike a fair balance between the author's right to control the dissemination of his works and the public interest in fostering their widest dissemination."[48]

Madison's unequivocal "the public good *fully* . . . coincides with the claims of individuals," [emphasis added] which relied on the working of the copyright scheme to achieve the widest and most efficient dissemination of an author's work—that is, a scheme in which costs and benefits are already balanced—has been replaced by a formulation that assumes that instead of the copyright scheme being *itself* in the public interest, there are deep conflicting interests between author and society requiring fundamental adjustments in the copyright scheme by Congress. Since the examination of the question underlying the assumption is one of the principal tasks of Congress when it approaches copyright revision, there might be grounds for worry that the question is answered before the inquiry is begun. For the fact is that the Register's implication—"*Accordingly,* the United States copyright law has imposed certain limitations and conditions on copyright protection" [emphasis added],—that copyright generally, and therefore existing law specifically, has had typically to adjust the internal workings of the copyright scheme because of deep conflicts of interest *is simply not borne out by the facts he cites:* the last three of his four "limitations and conditions" have merely to do with the technical trappings of the law or with the termination period in which the created work is to respond to the internal dynamics of the scheme; and the first, which is the only one that has to do with the internal logic and tensions of the scheme, refers merely to the traditional fair uses and to the three exceptions, essentially undisturbing to the copyright design, discussed above. In short, the Register has had to labor to establish what is on its face a straw man: there is no demonstration that the ordinary cost and access burdens that have from the outset characterized the exclusive-rights basis of copyright were mis-

---

[48] *Id.* at 6.

placed, and the "many situations" of conflict are uncatalogued. The copyright scheme has not, in fact, been characterized by "limitations and conditions" that control the fundamental workings of copyright. It has, on the contrary, been left to work in accordance with its internal dynamics, essentially undisturbed by statutory exemptions.

And this history is in accordance with the fundamental conceptualization of copyright, which is here obscured—namely, that *the copyright scheme, once a work is brought within its scope, is not thenceforward concerned with the internal reallocation of costs between creators and users.* Fair use, which is an integral part of that design, is therefore not either. Congress, however, *is* concerned with the allocation of costs, and reallocating them *is* its concern in making exemptions from copyright.

It would appear, therefore, that in posing the matter of the "exclusive right" tension by simply pitting authors' interests against society's, the reason for the need for changes, if any, in the copyright act is in advance of fact and analysis anticipated. Instead of the question being posed as dual—"What *technical adjustments* or *refinements of definition* are needed to permit the copyright scheme to respond to the internal dynamics set in motion by the initial public-policy risk taken by the state's reliance on copyright?" and "What *substantive policy adjustments* are needed so as to remove certain competing constitutional interests from copyright protection altogether?"—the task is seen rather as consisting almost wholly of the second. In a sense, therefore, it might be said that the wrong question has been posed and the proper answers therefore prejudiced. If in our analysis we find some difficulties with the congressional solution, the reasons may perhaps be found in this conclusory posture.

In dealing with infringement cases, the only line a court need draw is that between fair use and protected use. Once a legislature introduces the notion of exemptions, however, *two* lines must be drawn, as shown on Figure 1, among what are now three categories—fair use, exempted use, and protected use. The coherence of any such scheme will necessarily reflect the precision with which lines are drawn between fair use and protected use and between exempted use and fair use; and the second line will depend heavily on the first. Since the line between fair use and protected use has heretofore been drawn only by the courts, legislative precision will depend both on what the courts have in fact done and on what the legislature sees as having been done.

Our approaches to fair use and to exemptions in the 1976 act will necessarily be very different, for we are handicapped by the fact that Congress has given us no help with the first line. As we shall see, it

Figure 1. Schematic Rendering of the
Copyright Design

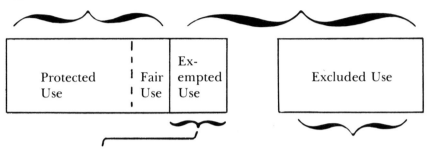

Protected by Copyright:
Costs accommodated within the
internal workings of the copyright scheme

Not Protected by Copyright:
Costs accommodated outside the
economics of copyright

Expressly exempted by statute
from the internal workings of
the copyright scheme.

Not within the scope of the
copyright scheme.

has failed to articulate a coherent rationale for copyright, it has failed
to define fair use, it has introduced confusions between fair use and
exempted use, and it has in the end tossed the fair use question, now
thoroughly enmeshed in contradictions, back to the courts. Before we
consider the complex matter of exemptions, therefore, we must try to
sharpen the notion of fair use—sharpen it at least enough to let us see
where a reasonable line might be drawn between fair use and exempted use. In the course of this exercise we shall suggest a more precise
fair use standard and test it against court decisions made in various
recent fair use cases.

II.

The Elements of Fair Use

A. The Section 107 Approach

If the purpose of a statutory definition of fair use is to articulate a coherent rationale for public policy, to indicate the general principles that follow from such a policy, and to establish or to refine a standard that will help courts in dealing with particular determinations of what they have long agreed, along with Judge Learned Hand, are "the most troublesome in the whole law of copyright,"[49] then the treatment of the issue in the new Copyright Act is very nearly a total loss. Section 107, which deals with fair use, reads as follows:

> *107. Limitations on exclusive rights: Fair use*
> *Notwithstanding the provisions of section 106, the fair use of a copyrighted work, including such use by reproduction in copies or phonorecords or by any other means specified by that section, for purposes such as criticism, comment, news reporting, teaching (including multiple copies for classroom use), scholarship, or research, is not an*

---

[49] Dellar v. Samuel Goldwyn, Inc., 104 F.2d 661 (2d Cir. 1939). Justice Story, the principal Supreme Court architect of early copyright adjudication, comments in the first fair use case: "This is one of those intricate and embarrassing questions in which it is not. . .easy to arrive at any satisfactory conclusion, or to lay down any general principles applicable to all cases . . .the lines. . .sometimes become almost evanescent, or melt into each other." Folsom v. Marsh, 9 Fed. Cas. 342, No. 4,901 (1841). And a more recent judge adds: "The doctrine is. . .so flexible as virtually to defy definition." Time, Inc. v. Bernard Geis Associates, 293 F. Supp. 130 (S.D. N.Y. 1968).

infringement of copyright. In determining whether the use made of a work in any particular case is a fair use the factors to be considered shall include:

(1) the purpose and character of the use, including whether such use is of a commercial nature or is for nonprofit educational purposes:

(2) the nature of the copyrighted work;

(3) the amount and substantiality of the portion used in relation to the copyrighted work as a whole; and

(4) the effect of the use upon the potential market for or value of the copyrighted work.[50]

The section has three serious defects. First, it does not attempt a definition of fair use at all. Second, by not providing the slightest guidance in the ordering of priorities in the application of the four "factors to be considered" it has not only said nothing not obvious about fair use, but, worse, implied that there *is* no general order of priority deriving from the copyright scheme. Third, by listing along with universally acknowledged examples of fair use (criticism, comment, and news reporting) those expansive and ambiguous uses (teaching, scholarship, research) that have raised issues having to do with significant exemptions from copyright, expressly dealt with as such in various ways in the statute, it thoroughly muddies the distinction between fair use and exempted uses.

The decision not to provide a definition of fair use in the statute is expressly set out in both the Senate and the House Committee Reports in the same words, in two separated paragraphs, as follows,

"Although the courts have considered and ruled upon the fair use doctrine over and over again, no real definition of the concept has ever emerged. Indeed, since the doctrine is an equitable rule of reason, no generally applicable definition is possible, and each case raising the question must be decided on its own facts. On the other hand, the courts have evolved a set of criteria which, though in no case definitive or determinative, provide some gauge for balancing the equities. These criteria have been stated in various ways, but essentially they can all be reduced to the four standards which . . . have been adopted in section 107. . . .

"The statement of the fair use doctrine in section 107 offers some guidance to users in determining when the principles of the doctrine apply. However, the endless variety of situ-

---

[50] Copyright Act, § 107.

ations and combinations of circumstances that can rise in particular cases precludes the formulation of exact rules in the statute. The bill endorses the purpose and general scope of the judicial doctrine of fair use, as outlined earlier in this report, but there is no disposition to freeze the doctrine in the statute, especially during a period of rapid technological change. Beyond a very broad statutory explanation of what fair use is and some of the criteria applicable to it, the courts must be free to adapt the doctrine to particular situations on a case-by-case basis. Section 107 is intended to restate the present judicial doctrine of fair use, not to change, narrow, or enlarge it in any way."[51]

Two characteristics of these paragraphs are particularly important for our purposes.

The first has to do with the diffident nature of this characterization both of the fair use problem and of the way Congress has decided to deal with it: that the concept has not really been defined by the courts, and that Congress does not attempt it either; that the several factors that are to be "included" in considering a particular case constitute "some" gauge to courts for balancing the equities and "some" guidance to users; and that the fair use doctrine *as stated in the whole of section 107* is not intended to change, narrow, or enlarge the "present judicial doctrine" in any way.

The second has to do with differences between the Senate and House Committee Reports in the general thrust of the text between and following the two paragraphs quoted here. The Senate version is consistent with these two quoted paragraphs, principally in considering the first sentence of section 107 as consisting essentially of some nonexclusive examples. For example, the Senate Committee Report in its text between the two quoted paragraphs says that "whether *a use* [emphasis added] referred to in the first sentence of section 107 is a fair use in a particular case will depend upon the application of the determinative factors, including those mentioned in the second sentence,"[52] and in an extended section following the second paragraph discusses the "four factors" in great detail.[53]

The House version, however, clearly uneasy with the failure to define fair use, attempts by an arbitrary recharacterization of the first sentence of section 107 to promote it to such a definition. "These

---

[51] Senate Committee Report, 62; House Committee Report, 65 and 66.
[52] Senate Committee Report, 62.
[53] *Id.* at 63-65.

criteria," it says, "are relevant in determining whether the basic doctrine of fair use, *as stated in the first sentence of section 107*, applies in a particular case."[54] The simple examples of "use" in the first sentence (Senate characterization) has in the House version become a "statement" of the "basic doctrine of fair use," something not even pretended to by the House text itself for the *whole* of section 107.[55] And perhaps to render this boot-strap upgrading of the first sentence less obviously without logical support, the House Report downgrades the long discussion of the "four factors" (which it acknowledges was, in its essentials, in its own 1967 report) by saying "it still has value" and then, perhaps because it belatedly sees problems it is reluctant to try to untangle at this stage of things, by omitting it altogether from the House Committee Report.[56] The House's insistence that the first sentence of section 107 somehow "states" the fair use doctrine comes close to being the "staccato bleat" that fair use is fair use characterized by Professor Kaplan.[57]

The principal reason for Congress' failure really to define fair use, to order the factors in some coherent way, or to draw a clearer line between fair use and exempted uses, as the legislative history of its gestation makes clear, is the disorienting impact of photocopying and phonorecording technology.[58] The advent of such technology, introducing a new and unsettling dimension into the whole copyright scheme, required of Congress a reexamination of fundamental copyright principles, a careful analysis in general terms of the internal dynamics of the copyright mechanism, the making of distinctions among the various elements to be considered, and the ordering of these considerations in a coherent way. What was needed was a sharpening of the concept of fair use, a narrowing of the definition so that it could with more precision be applied to particular cases by the courts, to whom the problem was expressly returned. Instead, almost the entire attention of Congress with respect to fair use was devoted to one aspect of the technical problem of photocopying, and the complex issues having in general to do with fair use were focused solely on the resolution of a single case—educational copying of copyrighted works.[59] That is, instead of facing squarely the primary question "What do we mean by fair use?" or the secondary question "How does

---

[54] House Committee Report, 65. [Emphasis added.]
[55] *See* text at n. 51, *supra*.
[56] House Committee Report, 67.
[57] *See* n. 69, *infra*.
[58] *See* n. 59 *and* text at note 77, *infra*.
[59] *See* Register of Copyrights, Supplementary Report on the General Revision of the U.S. Copyright Law, 89th Cong., 1st Sess., Copyright Law Revision,

the advent of the new technologies affect the conceptualization, and therefore the application, of the fair use doctrine?" Congress dealt with fair use on a tertiary level: "How do we fashion a fair-use statute so as to solve, by means of a compromise, a particular and expressly formulated exemption from copyright, the photocopying reproduction of copyrighted works for educational purposes?"[60]

The consequence was an utter dilution, by all parties to the public debate, of the notion of fair use, which more often than not came to be used merely to mean free use in the context of a discussion of a particular *exemption* from copyright. In the process, the line between fair use and possible exempted use was systematically obscured. The treatment of the question in the legislative history makes this clear. Here, for example, is a part of the Senate Committee Report not retained in the House Report: "However, since this section [i.e. § 107] will represent the first statutory recognition of the doctrine in our copyright law, some explanation of the considerations behind the language used in the four criteria is advisable. This is particularly true as to cases of copying by teachers, and by public libraries, since in these areas there are few if any judicial guidelines."[61] That is, an acknowledged exempted use, such as library copying, dealt with expressly as

---

Part 6 (House Judiciary Comm. Print, 1965), where the history of the drafting of the fair use section is set out.

The Report recounts the development of two earlier drafts of the fair use section more or less like that now in the statute, but concludes, at 28:

"Since it appeared impossible to reach agreement on a general statement expressing the scope of the fair use doctrine, and since in any event the doctrine emerges from a body of judicial precedent and not from the statute, we decided with some regret to reduce the fair use section to its barest essentials. Section 107 of the 1965 bill therefore provides:

Notwithstanding the provisions of section 106, the fair use of a copyrighted work is not an infringement of copyright."

The Register, in suggesting a bare statement that the doctrine of fair use exists, thus accepted what he had theretofore not—the unanimous recommendation of the nine copyright specialists asked by the Copyright Office to comment on the fair-use study prepared for Congress in 1958, who urged either that no mention be made or that no definition of the doctrine of fair use be attempted in the statute. *See* Latman, *Fair Use of Copyrighted Works*, in U.S. Copyright Office, Copyright Law Revision, Study No. 14, 86th Cong., 2d Sess. 39-44 (Senate Judiciary Comm. Print 1960) [hereafter cited as "Latman"].

The House was not persuaded by the Register's arguments and in 1967 adopted the approach and the language still substantially in section 107 of the new Copyright Act.

[60] *See* text at n. 77, *infra*.

[61] Senate Committee Report, 62.

an *exemption* in *another* section, 108, is here lumped in with fair-use analysis. The House version by omitting this section of the Senate text seems implicitly to recognize the contradiction in it and deals with it as it did with the failure of section 107 actually to define what fair use was,[62] namely, by redefining the problem. It simply says that "a special exemption freeing certain reproductions of copyrighted works for educational and scholarly purposes from copyright control is not justified."[63]

In this circumstance, in order ultimately to deal with how the exemptions in the new act affect the copyright scheme, we must start afresh. Our task is twofold—to disentangle exempted-use notions from fair-use notions, and to fashion a general but lean statement of fair use rationale from which courts might apply equitable principles to particular cases. We embark first on disentanglement, an exercise that will give us some surprising assists in approaching the definition of fair use itself.

B. The Traditional Fair Use Configuration

The Register of Copyrights, in his 1961 Report orienting Congress on the issues to be dealt with in copyright revision, lists, after a

---

[62] *See* text at n. 54, *supra.*

[63] House Committee Report, 67.

That there is a line to be drawn between fair use and exempted use is sometimes seen and sometimes obscured in the testimony by the Ad Hoc Committee of some forty-one educational institutions in the 1973 Senate Committee hearings of section 107. That testimony called for both a "clear delineation of 'fair use'" and "in addition" an "educational exemption," but the discussion of both is phrased throughout in fair use terms. In any event, it is clear that the Committee, in combining fair use delineation with school-use exemptions in section 107, did not see the distinction as important or useful. Senate Committee on the Judiciary, Hearings before the Subcommittee on Patents, Trademarks, and Copyrights, S. 1361, 93d Cong., 1st Sess. (1973), 180-85.

Similarly, the distinction is not rendered clearer by the testimony on behalf of the Association of American Law Schools, the American Association of University Professors, and the American Council on Education. Though protesting that he does "not seek to remove protected material from copyright control," the spokesman argued that "educational use" should go beyond the traditional fair use "precedents of the past" because teaching and research require "access to the work product of allied researchers" and "it is not possible for every university and law library to acquire one or more copies of every book needed for research and teaching." He urged that express exemptions be made for "research and teaching" under the fair use doctrine either in the statute or in the legislative history of the section on fair use. *Id.* at 203.

brief introduction to the notion of fair use, examples of "the kinds of uses that may be permitted under that concept:

> "—Quotation of excerpts in a review or criticism for purposes of illustration or comment.
> "—Quotation of short passages in a scholarly or technical work, for illustration or clarification of the author's observations.
> "—Use in a parody of some of the content of the work parodied.
> "—Summary of an address or article, with brief quotations, in a news report.
> "—Reproduction by a library of a portion of a work to replace part of a damaged copy.
> "—Reproduction by a teacher or student of a small part of a work to illustrate a lesson.
> "—Reproduction of a work in legislative or judicial proceedings or reports.
> "—Incidental or fortuitous reproduction, in a newsreel or broadcast, of a work located at the scene of an event being reported."[64]

Now, an essential characteristic of this list is that it does not, on the whole, deal at all with photocopying as we know it. And aside from the rather special case of repairing a damaged copy, only one of the eight examples—reproduction of material to illustrate a lesson—has to do with the copying of a copyrighted work *for its own sake,* and even that is narrowly defined. The list, casual or studied as it may be, reflects what in fact the subject matter of fair use has in the history of its adjudication consisted in: *it has always had to do with the use by a second author of a first author's work.* Fair use has not heretofore had to do with the mere reproduction of a work in order to use it for its intrinsic purpose—to make what might be called the "ordinary" use of it. When copies are made for the work's "ordinary" purposes, ordinary *infringement* has customarily been triggered, not notions of fair use. In the special Copyright Revision study on fair use prepared by the Copyright Office for Congress,[65] not a single case cited holds that straightforward reproduction of a copyrighted work for its own sake constitutes fair use.[66] The "reproduction" of a work for its intrinsic

---

[64] Report of Register, *supra* n. 33 at 24.

[65] Latman, *supra* n. 59.

[66] There were in fact no cases cited in which the fair use issue, when it was

use was first posed in fair use terms by the recent *Williams and Wilkins*[67] case, in which copies of articles in specialized medical journals were systematically, freely, and widely distributed by centralized government duplicating services to the specialist users for which they were in the first instance written.[68] There would appear, then, to have heretofore been a tacit distinction between the kind of infringement that has to do with the "unfair use" of a prior work in a later work and the kind of infringement that has simply to do with making or multiplying copies of an existing work.[69] (Congress, in overruling *Wil-*

---

raised, had to do simply with reproduction of an unmodified copyrighted work for its own sake. The cases usually cited—*e.g.,* in Williams and Wilkins v. United States, 487 F.2d 1345, 1353, 1377 (Ct. Cl. 1973)—all have to do with some sort of adaptation or modification of a work: in Hill v. Whalen & Martell, Inc., 220 Fed. 359 (S.D. N.Y. 1914), the "Mutt" and "Jeff" cartoon characters were used in a stage production; in Leon v. Pacific Telephone and Telegraph, 91 F.2d 484 (9th Cir. 1937), a telephone directory was rearranged [*see* text at note 115, *infra*]; in Folsom v. Marsh, 9 Fed. Cas. 342 (D. Mass 1841), a large-scale abridgement and adaptation of a book was at issue; in Public Affairs Associates, Inc. v. Rickover, 284 F.2d 262 (D.C. Cir. 1960), separate speeches had been collected into a new work and published; in Wihtol v. Crow, 309 F.2d 777 (8th Cir. 1962), a teacher reproduced copies of his own arrangement of a copyrighted song.

[67] 420 U.S. 376, 95 S. Ct. 1344 (1975), aff'g, by an equally divided court, 487 F.2d 1345 (Ct. Cl. 1973).

[68] "What we have before us is a case of wholesale, machine copying, and distribution of copyrighted material by defendant's libraries on a scale so vast that it dwarfs the output of many small publishing companies. In order to fill requests for copies of journals, the [National Institute of Health] made 86,000 Xerox copies in 1970, constituting 930,000 pages. In 1968, the [National Library of Medicine] distributed 120,000 copies of such journal articles, totalling about 1.2 million pages." *Williams and Wilkins,* 487 F.2d 1345 (Ct. Cl. 1973), at 1364, Judge Cowen, in dissent.

[69] Professor Goldstein's broad formulation ("The effect of the fair use defense is to excuse otherwise infringing conduct in circumstances where the public interest compels free access." Goldstein, *Copyright and the First Amendment,* 70 Col. Law Rev. 1011 (1970), by characterizing fair use as within the infringing ambit, appears to leave it undistinguished from exempted use. Occasionally, so does Professor Nimmer: "Congress may, at its discretion, limit the copyright monopoly under the doctrine of fair use." Nimmer, *Does Copyright Abridge the First Amendment Guarantees of Free Speech and Press?* 17 U.C.L.A. Law Rev. 1204 (1970).

Stated thus unmodified, this notion of fair use as excusing an infringement would appear to apply to intrinsic-use copying of the sort represented by photocopying, something not necessarily meant by other authorities in discussing fair use, though it is sometimes seen that way. For example, Professor Kaplan cites Nimmer, section 145, as taking "fair use to refer to a set of justifications averting liability for what on the face of things is infringement." B. Kaplan, An Unhurried View of Copyright

liams and Wilkins by expressly characterizing in section 108 of the new
Copyright Act as an exemption what the court in that case called fair
use, in fact does what the committee reports both say is not being
done, namely changing the "present judicial doctrine . . . in any
way.")[70]

But there is a point at which the separate uses made by such a
distinction meet, and it is there that the seed of conceptual confusion
is to be found. That point is the copying, traditionally by hand (and
later by typewriter), by a private reader, scholar, writer, student, or
teacher, of a work for the copier's own private use.[71] It is this "copying for private use" that is at the crossroads of traditional fair use notions and the intrinsic-use questions of infringement posed by photocopying. According to Professor Nimmer, there has never been a "reported case on the question of whether a single handwritten copy of
all or substantially all of a protected work made for the copier's own
private use is an infringement or fair use."[72] Yet it is precisely here,
where there has heretofore been no question disturbing of either the
basic rationale or the operative dynamics of the copyright scheme,
that the disorienting question posed by photocopying has arisen. The
reason, it is clear, is that of all uses traditionally accepted as "fair," this
one alone has to do with copying for its own sake. Those aspects of
the copyright scheme that in consequence require reconciliation we
deal with below, following a consideration of fair use standards. Here,

---

67. But Professor Nimmer's remark is in the context not of intrinsic-use
reproduction, but of the notion of "substantial similarity"—i.e., the use by a
second author of a first author's work. This blurring of the intrinsic-use
notion allows Professor Kaplan to go on to say that "it would, I think, be
possible to dispense with [fair use as a separate analytical instrument]", *id.*
at 67-68 [*see* n. 82 *infra*], though when he later addresses the difficult problem of photocopying, he seems to criticize the then (1964) draft of the
Copyright Revision Bill for failing to define fair use: "The Revision Bill is
reduced to uttering on the subject only the staccato bleat of just the words
'fair use.'" *Id.* at 102.

[70] *See* text following n. 51, *supra.*

[71] The "hand-copying" analogy was the point of departure for the so-called
"gentlemen's agreement" issued in 1935 by an association of book publishers and an informal committee of research-and-education-oriented organizations. Quoted in the *Williams and Wilkins* decision and in 1 J.
DOCUMENTARY REPRODUCTION 29 (1939). For a perspective on the
technological aspects of the "agreement," *see* n. 94, *infra. See* Appendix E,
infra.

[72] NIMMER ON COPYRIGHT, § 145, at 656.3. Professor Nimmer cites in a foot-
note the remark in *Williams and Wilkins:* ". . .it is almost unanimously accepted that a scholar can make a handwritten copy of an entire copyrighted article for his own use. . ."

we concern ourselves only with disentangling exempted-use issues from Congress's single-minded photocopying treatment of fair use.

C. The Confusion of Photocopying Exemptions with Fair Use

The fair use problem was in fact first posed in photocopying terms by the Register in his 1961 Report. The sentence preceding that introducing the list of examples quoted above[73] reads: "[Fair use] eludes precise definition; broadly speaking, it means that a reasonable portion of a copyrighted work may be *reproduced* without permission when necessary for a legitimate purpose which is not competitive with the copyright owner's market for his work."[74]

It is perhaps not too much to say that a central question with respect to fair use has been begged when "use" is identified solely in terms of "reproduction." And this conceptualization of it was at once reinforced, in the section of the Report immediately following, by the invoking of the notion of fair use in connection with the problem of library photocopying,[75] a matter subsequently partly disentangled from fair use notions by being dealt with as an express exemption in a different part of the statute.[76] And though the Register was careful to try to distinguish between the thought that the fair use doctrine might have *some* application to library phtocopying and the thought that photocopying, having dimensional aspects similar to printing itself, raised questions that went beyond the notion of fair use, the damage was done. The term fair use was thenceforth used, either consistently or occasionally, by all parties to the public debate about specific exemptions as if it meant simply free use by means of reproduction of a work. Its doctrinal content became nearly extinguished, and Congress dealt with the issues posed indiscriminately either as separate exemption sections of the statute or as expansive readings, set out in the commentary on the bill, of fair use.

Thus, library copying was dealt with exhaustively and in great detail as an express exemption, section 108, while educational photocopying, which in some respects allows more extensive and greater multiple copying than the library exemption, has been dealt with essentially in the committee reports on the fair use section. Yet in other parts of the statute Congress saw the complexity and magnitude of educational use as requiring express and narrowly defined statutory

---

[73] *See* text at n. 64, *supra.*
[74] Report of Register, *supra* n. 33 at 24. [Emphasis added.]
[75] *Id.* at 25.
[76] § 108. *See* text at n. 70, *supra.*

exemptions. For example, section 110 goes to great lengths to spell out how copyrighted works may be "performed" or "displayed" in educational situations by teachers, and section 111 extends those rights to electronic retransmissions. Thus Congress has defined both an express *educational* use (performance and display) and an analogous *photocopying* use (library copying, which may of course be for "teaching, scholarship, or research") as uses requiring exemption treatment. It is difficult to reconcile this with the different treatment of *educational photocopying,* embodying both issues.

In any case, by the time the House subcommittee came to consider the 1975 bill passed by the Senate, the entire 32-page orienting background chapter on fair use of the Register's report to the House committee,[77] except for one paragraph, dealt solely with the question of educational exemptions. If there was in fact a conceptual fair use question that still needed careful examination, there was no sign of it here. And Congress, seeing its task at this point as brokering a private agreement among educators, authors, and publishers, was in no mood to reexamine general principles.

But we cannot leave matters thus if we are to deal with the exempted-use issues in a coherent way. We shall examine educational photocopying in Chapter III,[78] in the context of the exemptions to which, we shall argue, it is more akin. Here we must next make the distinction that Congress obscured between the accommodation of relationships *within* the copyright scheme (fair use), and therefore appropriately the province of the courts, and the balancing of interests *outside* the copyright scheme (exempted use), and therefore the province of Congress.

D. A Dual-Risk Approach to Fair Use Analysis

Leaving aside the first sentence of section 107,[79] which by the unevenness of the examples listed raises more questions than it answers, we are left with what amounts to Congress' instruction to the courts simply to take account of all considerations inherent in the copyright scheme, though Congress is not consistent about what the courts might consider. Both legislative committee reports say that the criteria essentially "can *all* be reduced to the four standards . . . adopted in

---

[77] Second Supplementary Report of the Register of Copyrights, October-December 1975, chapter I1.
[78] *See* text at notes 203-210.
[79] *See* § 107 at n. 50, *supra.*

section 107," [emphasis added] something that is not quite congruent with the statute's use in section 107 of the word "include," defined in section 101 of the statute as "illustrative and not limitative."[80] But however that may be, the "four factors," in the accidental order in which they have been repeated by courts, commentators, and reports, have become a litany, as they have been for Congress, obscuring their discrete meanings and forestalling their arrangement in a sequence reflecting what might be seen as the logical priorities of the copyright scheme. We shall in due course come back to them and try to order them in a more useful way, but we must first look once more at certain aspects of the constitutional configuration of the copyright scheme.

Again, the copyright scheme is itself not fundamentally concerned with the *allocation* of costs: as we have seen in Chapter I, it assumes that the workings of the scheme will take care of that in accordance with the balancing taken by the initial risk. It *is* concerned, however, with whether there is in fact a cost at all—and that, in the first instance, is what fair use involves. Fair use, which the copyright scheme implicitly incorporates, has to do with whether a particular cost-free use is one both foreseen by the author and contemplated by the Constitution. As we shall in due course see,[81] the courts have had most difficulty in fair use adjudication when they have confused the question of the allocation of costs with that of whether the free use was one the first author should have known he was risking. The second question of course turns on the implied meaning of the term "exclusive right," and it is to a reexamination of that complex constitutional term that we return.

It is clear, of course, that no one—least of all the author—means the phrase literally. What the author fashions out of his intellect and sensibilities he *expects* to be used by other minds and other sensibilities. That is why he does it. He hopes that people will recite his poems, that other thinkers will cite his work and rely on it, that students will learn from him, that the world will take note of what he has wrought, and that the private reader will copy out his words and sing his songs. And for such use he expects neither to be asked nor to be paid.

But somewhere shortly beyond that he has economic expectations appropriately deriving from what society offered him in the copyright scheme. Similarly, society does not intend that the "exclusive right" language shall bar appropriate use of his work by others in the furtherance of progress of knowledge and the arts. It is at the junction of

---

[80] *See* text at n. 51, *supra*.
[81] *See* Part F of this Chapter, *infra*.

these two sets of expectations—about costs and about access—that the question of fair use arises. The appropriateness of each must necessarily derive from one's perspective. If society is to rely on the copyright scheme for the allocation of costs, then at the margin it must see its interests in adjudication as being balanced more on the side of access. If the author is to rely on his control of access to reward him economically, he will see his interests in adjudication as being balanced more on the side of payment.

Accordingly, the court's notion of fairness in the use of copyrighted materials will proceed from this dual perspective about "normal expectations": the author expects that the copyright scheme itself will sometimes require use of his work necessary in the public interest for which he will not be paid, and society expects that the copyright scheme will either allow such use without reducing the author's incentive or impose no excessive burdens on the public when use is controlled. Put another way, in entrusting their respective interests to the copyright scheme, author and society each takes a risk that the costs to each would not be unacceptable. In a particular instance, their dual risk might be posed by a pair of questions: Is this use within the risk the author was taking that he would not be paid? Is this use within the risk society was taking that the author would assert control of access? Since both questions turn on the appropriate expectations of each, the determination of fair use in a particular instance will decide whether the author's expectation of economic reward was or was not appropriate, and such a determination ought to coincide with a simultaneous judgment about whether society's expectation of denial of access was or was not appropriate.[82]

---

[82] Though Professor Kaplan, in AN UNHURRIED VIEW OF COPYRIGHT (1966), suggests that "it would. . .be possible to dispense with [fair use as a separate analytical instrument]," at 68, he employs a similar "expectation" concept in describing Lord Mansfield's decision in Sayre v. Moore, 1 East 361n., 102 Eng. Rep. 139n. (K.B. 1785): "even a bodily taking. . .could be defended if it was conceived not to interfere unduly with the normal economic exploitation of the copyright," Kaplan, at 17.

Professor Nimmer's formulation has a different emphasis: "It is believed that the actual decisions bearing upon fair use, if not always their stated rationale, can best be explained by looking to the central question of whether the defendant's work tends to diminish or prejudice the potential sale of the plaintiff's work." NIMMER ON COPYRIGHT at 646. This echoes the opinion, for example, in the "Mutt and Jeff" case: "One test which, when applicable, would seem ordinarily decisive, is whether or not so much has been reproduced as will materially reduce the demand for the original." Hill v. Whalen & Martell Inc., 220 Fed. 359 (S.D.N.Y. 1914). But what or-

That, in the end, is solely what fair use is, and the statute ought simply to say so. Section 107 does not. Thus, we might begin to construct a more useful fair use statute by including both notions:

"Fair use is use that is necessary for the furtherance of knowledge, literature, and the arts AND does not deprive the creator of the work of an appropriately expected economic reward."[83]

There is nothing novel about this notion. Most attempts at defining the controlling factor in particular decisions, or in decisions in general, cautiously conclude with an economic formulation more or less like this one. What is new is that it would be expressed in the statute rather than sidled up to. It simply makes explicit the basic economic risk taken by the constitutional copyright clause and sets out without blinking the controlling relationship between the two essential criteria that Congress has failed to face up to in the text of the statute. But there is no dodging these matters: there is no way of avoiding the danger of the expansiveness of a term like "necessary," and no qualification of the word can help. Clarity requires *in the statute itself* an express statement of the author's economic incentive. The fierce debate about photocopying exemptions that the Register of Copyrights recounts[84] as having taken place under the nominal rubric of fair use is not resolved by Congress' gingerly avoiding a statement of the proper relationship of the internal factors that balance copyright. Having failed to state it in a preamble of purpose, Congress ought at least to do it here, in the fair use context.

It is of course clear that the key phrase in the suggestion proposed here is "an appropriately expected economic reward." Under

---

dinarily happens in copying of this sort is not that the use has "materially reduced the demand for the original," but, more precisely, that the "normal expectations" of the author of the original work have been disappointed.

The normal expectations of an author will of course be different at different times under the influence of court decisions and congressional actions, as well as of technological and economic change.

[83] There is in this formulation something of the "implied consent" notion suggested in R. C. DeWolf, AN OUTLINE OF COPYRIGHT LAW 143 (1925), though his expression of it—that fair use is a use "technically forbidden by the law"—does not in terms distinguish exempted use from fair use.

Though Latman says the consent theory "fails as an overall basis of fair use," *supra* n. 59 at 31, he nevertheless finds it the most useful way in which to sum up the criteria for fair use: "the tests may perhaps be summarized by: importance of the material copied. . .from the point of view of the reasonable copyright owner. In other words, would the reasonable owner have consented to the use?" *Id.* at 15.

[84] *See* n. 59, *supra*.

such a formulation, it would be the task of the court to decide, bearing in mind the dual risk embodied in copyright, whether the author from his initial position ought, in view of the intellectual nature of his work and of the consequent need for others to "use" it, to have expected *both* to control the sort of subsequent use in question *and* to have been paid for it. If the answer is no, the courts would find the use fair and within the copyright scheme. If the answer were yes, the courts would protect the author's copyright interest and defer to Congress on the question of whether a reallocation of costs requires an exemption from copyright.

Have we thus merely begged the question? Not if notice considerations have any use in the formulation of a statute: such a definition in the law would give appropriate notice to the creator of the work and to the second user alike of the normal relationship between the two elements that the copyright scheme holds in tension. And perhaps more important, as we shall see when it is applied to particular cases, a definition that makes that relationship explicit is the necessary bedrock for the ordering of the separate factors that courts must consider in reaching particular judgments.

In turning to those "four factors" we find, in fact, that the logic of our formulation, whatever its persuasiveness, has already engaged us in suggesting priorities among the four factors.[85] In fashioning a statement embodying the essential economic configuration of the copyright scheme, we have had to find of *first* importance the factor that the statute puts *last*. That factor is mislocated in section 107 of the Act perhaps in proportion to the degree in which our analysis is persuasive. But we need not look far for support for the economic orientation we have taken. The "four factors to be considered" are wholly adumbrated, if not fully expressed, by Justice Story in the first American fair use case, *Folsom v. Marsh:*

---

[85] Professor Nimmer, in NIMMER ON COPYRIGHT, 645-46, quotes as typically of "little assistance" the mere listing of the factors in Mathews Conveyor Co. v. Palmer Bee Co.: "the court must look to the nature and objects of the selections made, the quantity and value of the material used, and the degree in which the use may prejudice the sale, diminish the profits, or supersede the objects of the original work." 135 F.2d 73, 85 (6th Cir. 1943). And the court in Meredith Corporation v. Harper & Row, Publishers, Inc., after noting and quoting Nimmer's citation of *Mathews*, establishes its own order of priority: "In determining whether the use here is 'fair,' I conclude the following three factors should be considered: (1) the competitive effect and function of the usage, (2) the quantity of the materials used, and (3) the purpose of the selections made." 378 F. Supp. 686, 689 (S.D.N.Y. 1974).

"The question then is, whether this is a justifiable use of the original materials, such as the law recognizes as no infringement of the copyright of the plaintiffs. It is said, that the defendant has selected only such materials, as suited his own limited purpose as a biographer. That is, doubtless, true; and he has produced an exceedingly valuable book. But that is no answer to the difficulty. It is certainly not necessary, to constitute an invasion of copyright, that the whole of a work should be copied, or even a large portion of it, in form or in substance. If so much is taken, that the value of the original is sensibly diminished, or the labors of the original author are substantially to an injurious extent appropriated by another, that is sufficient, in point of law, to constitute a piracy pro tanto. The entirety of the copyright is the property of the author; and it is no defence, that another person has appropriated a part, and not the whole, of any property. Neither does it necessarily depend upon the quantity taken, whether it is an infringement of the copyright or not. It is often affected by other considerations, the value of the materials taken, and the importance of it to the sale of the original work . . . . In short, we must often, in deciding questions of this sort, look to the nature and objects of the selections made, the quantity and value of the materials used, and the degree in which the use may prejudice the sale, or diminish the profits, or supersede the objects, of the original work."[86]

Now the central characteristic of the factors as set out in *Folsom* is their generally *economic* orientation, with Justice Story in the end expressly invoking the economic interests of the first author in a three-phrase peroration on the same theme. As we shall see, the courts have on the whole followed this formulation and have got into difficulties only when they have separated the "nature and objects of the selections made" from the economic orientation.

That orientation ought to help us in ordering the other factors. If we take our perspective from the first author's risk, then the nature of his work must color his expectations about how it will be used by a second author, and the "purpose" and "character" of the use must in turn color the second author's expectations, which embodies society's. The first author will know, for example, that for an informational work such as a biography or compilation, there will be expectations different from those attendant on a piece of music or a work of fic-

---

[86] 9 Fed. Cas. 342 (No. 4901), 27. (C.C. D. Mass. 1841).

tion.[87] "The nature of the copyrighted work" will govern the normal expectations of both society and the first author, and on the appropriateness of those expectations will depend both the second author's sense of what use he can make of the first author's work and a court's determination, on society's behalf, of whether he was right. That is the essential fairness question: fairness turns on perspective, and this order of things, which begins with fairness to the particular first work and proceeds to the point of view of the general scheme, establishes that perspective.[88]

In that sense, the "substantial injury" test used often by the courts might be characterized as *retrospective*, and the test looking toward the

---

[87] *See* Gorman's seminal article, *Copyright Protection for the Collection and Representation of Facts*, 76 HARV. L. REV. 1569 (1963). [hereafter cited as "Gorman"].

[88] The court in Rosemont Enterprises, Inc. v. Random House, Inc. [*see* text, *infra,* at n. 106], takes its orientation from the second author: "Whether the privilege may justifiably be applied to particular materials turns initially on the nature of the materials, e.g., whether their distribution would serve the public interest in the free dissemination of information and whether their preparation requires some use of prior materials dealing with the same subject matters." 366 F.2d 303, 307 (2d Cir. 1966). This primacy of the *user's* interest as stated in *Rosemont* was cited and elevated to the governing factor in the Court of Claims decision in *Williams and Wilkins*, 487 F.2d at 1352.

But any definition of fair use that does not begin with the author's perspective and proceed to what from his initial position he ought appropriately to expect ends up either reasoning in circles or begging the fundamental question. Consider, for example, Professor Gorman's formulation that courts "should label as 'infringements' those works which interfere unduly with the monopoly of the copyright holder without bringing a commensurate benefit to the public, and as 'fair use' those works which interefere but slightly with the copyright monopoly while offering much to society." Gorman, *supra* n. 87, at 1604. The logic of his fair use test is that it is all right for the exclusive rights of the author to be encroached upon even "unduly" so long as the benefit to the public is equal to it. With that as a standard, it is scarcely necessary to get to his second test, that it is all right for the exclusive rights of the author to be encroached upon "slightly," so long as society is "much" benefited. Under either test, all that appears to count is the degree of benefit to society, with an emphasis far removed from any notion of initial risk by either.

That the more accurate perspective on fair use has somehow to do with the dual constitutional purpose to give incentive to the author for the benefit of society than with the unfettered "benefits to the users" test may be suggested if the fair use notion of the "right to comment", for example, were rephrased in risk-analysis terms: that authors have no right to expect to prevent comment.

author's fair and appropriate expectations, *prospective*. Courts often use both tests simultaneously. What is argued for here is that the prospective test ought to govern—that the retrospective test of "substantiality of copying" can have meaning only in the light of the appropriate expectations of author and society alike.[89]

We accordingly come to the last of the four factors set out in the statute, "the amount and substantiality of the portion used in relation to the copyrighted work as a whole," which appears to suggest a conceptually workable test of injury in a fair use context. But this wording in fact tries too hard, losing accuracy and making for both a redundancy and a begging of a question. A formulation that identifies a factor to be considered in the same neutral mode as the other factors would omit the word "substantiality": "the amount of the portion used in relation to the copyrighted work as a whole." It is a consideration of that relationship, along with the others, that might result in a finding of substantiality—but substantiality itself is the ultimate fact to be found. That is, if a use is substantial it cannot be fair use. A substantial taking is the definition of *infringement*. Any excusing of a substantial taking must be an exemption from copyright. Accordingly, a more accurate statement of the *factor* to be considered would simply be the "extent of the use." That serves as well the interest of leanness in the definition, not always perceived or kept separate in analysis, between three different uses of "substantiality," namely, (1) substantiality of the degree of similarity (i.e., the question of whether there has in fact been copying), (2) the substantiality of the copying (i.e., the question of how much has been copied), and (3) the substantiality of the economic deprivation of the first author (i.e., the question of the

---

[89] As Professor Kaplan points out, with respect to Dodsley v. Kinnersley, Amb. 403, 27 Eng. Rep. 270 (ch. 1761), having to do with an unauthorized abridgment of Johnson's *Rasselas* in a magazine, the court's dismissal of the case because the defendant had not interfered with the sale of the *book* suggests that "even a bodily taking. . .could be defended if it was conceived not to interfere unduly with the normal economic exploitation of the copyright." B. Kaplan, An Unhurried View of Copyright 17. Both the case and Professor Kaplan's comment suggest the central factor: that what would be the "normal exploitation" of a copyright can change—that an abridgment not seen as a normal exploitation of the copyright would today be seen as one of the expectations for which an appropriate economic reward would be within the expectations of both author and society alike. The view unifying *Dodsley* with *Rosemont* is that the author not be deprived of what he would otherwise normally expect to get. This allows for changes in how a work might be used "fairly," but the standard and rationale are the same. *See also* n. 119, *infra*.

appropriately expected economic reward).[90] Accordingly, the next sentence of the section might read:

"In determining whether the use made of a work in a particular case deprives its creator of such a reward, account should be taken first of the nature of the copyrighted work and then of the purpose, character, and extent of the use."

The effect of this formulation is to reorder the "four factors" of the fair use section in the sequence 4, 2, 1, 3, an order that reflects a coherent view of the nature of the copyright scheme.

And we need go no further. Courts applying a statute embodying that order of the factors ought to find fair use adjudication more manageable. The listing of examples, as we have seen, raises more questions than it answers,[91] obscuring the very distinction among the three sorts of uses a fair use definition ought to make clearer—(1) normal expectations of reward, (2) wholly accepted free use and access, and (3) the areas at the margin that cause difficulty in fair use adjudication. Even the invoking of examples not listed in the statute, but cited elsewhere by Congress in the Committee Reports, runs the danger of blurring distinctions. For example, the Senate Commitee Report comments: "With certain special exceptions (use in parodies or as evidence in court proceedings might be examples) a use that supplants any part of the normal market for a copyrighted work would ordinarily be considered an infringement."[92]

Two inaccuracies are embodied here. First, the relationship of fair use notions to the economics of copyright is confused by the juxtaposition of "supplants *any* part of the *normal* market" with "would *ordinarily* be considered," for what is impliedly being described is an *exemption*. Fair use, if it is to have any coherent meaning, must *by itself* suggest to both creator and society that it is *not* part of the normal market. Second, by thus asserting that uses in parodies and court proceedings (though "exceptions") are uses that supplant part of the "normal market," the sentence on the one hand obscures the utterly pure fair use character of copying in court proceedings and on the

---

[90] With respect to "substantiality," *cf.* NIMMER ON COPYRIGHT at 627-35.

[91] None of the nine consultants of the Copyright Office asked to comment on the appropriate statutory approach to fair use recommended listing any examples at all. Though Professor Nimmer, for example, recommended that the act "expressly adopt and codify the existing judicial doctrine of fair use," he warned that "such recognition should be in general terms, and should not attempt any specific enumeration of particular instances of fair use. Specific enumeration would be undesirable. . ." Latman, *supra* n. 59 at 42.

[92] Senate Committee Report, 65.

other hand finesses the complexities (and the consequent analytical care required) of "appropriate" use in parodies.[93]

Matters ought to be clearer with a fair use definition shorn of its ambiguous examples, incorporating an express definition that identifies the tensions inherent in the copyright scheme, and reordering the "four factors" in a sequence whose priorities reflect the fundamental relationship of the author-as-instrument and the society-at-risk.

E. The Intrinsic Use Question

Since we have developed the "appropriately expected economic reward" standard out of the traditional fair use matrix having to do with the second author's incorporation of a copyrighted work in his own work, there remains to consider how it accommodates the expansive pressures on the fair use notion exerted by the actual reproduction of a work made possible by ubiquitous photocopying and photorecording.

So long as the author could, because of technological constraints, control the reproduction of his work for its own sake, his expectations of economic reward and society's view of the appropriateness of his expectations when his work *was* reproduced pretty much coincided: reproduction in such circumstances either had to trigger infringement or to be seen as so minimal as not to raise fundamental questions.[94] So far as notions of fair use are concerned, then, the central question with respect to ubiquitous photocopying has to do with the shape of the basic dual risk.

Again, questions of access and of cost are at issue, but a coherent approach turns initially on making a distinction between two kinds of "access." The traditional fair use notion has had to do with the sort of access that the mind of a user has to a copyrighted work: the work is instantly accessible on sight or hearing, and the question is what the second user would or should be allowed to "do" with it—or, more accurately, what the creator of the work ought to have expected him freely to do with it. Use for photocopying and phonorecording in-

---

[93] *See* text at n. 107, *infra.*

[94] The limitations of photocopying technology in 1935 governed the central premise of the so-called "gentleman's agreement" of 1935, in which publishers' and authors' expectations were set out. *See* n. 71, *supra.* Much has been made of the point that in this "agreement" the copyright owners had by that instrument fully acknowledged the acceptability of reproducing copyrighted works in the library circumstances defined there. But a reading of the agreement indicates that the governing point is almost surely the

volves a different sort of access: the possibility—and the capability—of instant reproduction of the work in the same mode and for the same purpose the original was in the first place acquired. With the first sort of access and use we can deal under notions of fair use, even when it involves photocopying, for appropriate expectations of economic reward are essentially unchanged. The second kind of access and use triggers questions of reallocation of costs, however, and insofar as a use is of that sort, it is appropriately dealt with as an exemption from the normal workings of the copyright scheme, which we take up below, in Chapters III and IV.

—————

reverse: in 1935 the sort of copying being dealt with was either microfilm or the cumbersome, slow, and expensive reproducing mechanism, such as photostat, used in drafting rooms and graphic arts studios. That is, the sort of copying addressed there was, as with hand-copying, so minimal as not to raise fundamental questions: that is what the situation was as a practical matter, and it was implicitly defined as such in the "agreement." Chief Judge Cowen of the Court of Claims, in dissent in *Williams and Wilkins,* commented on the "agreement" as follows: "[T]he 'agreement' was drafted at a time when photocopying was relatively expensive and cumbersome; was used relatively little as a means of duplication and dissemination; and posed no substantial threat to the potential market for copyrighted works. Beginning about 1960, photocopying changed character. The introduction to the marketplace of the office copying machine made photocopying rapid, cheap, and readily available. . . .[T]he 'gentlemen's agreement' by its own terms condemned as 'not. . .fair' the making of photocopies which could serve in 'substitution' for the original work." 487 F.2d 1345, 1380 (Ct. Cl. 1973). On the central issue of the economic disturbance of copyright-scheme dynamics, Judge Cowen also quotes a 1968 study:
     "In a thorough and thoughtful discussion of the effect of reprography, prepared at the University of California, Los Angeles, and funded by the National Endowment for the Arts, it is stated:
          It has long been argued that copying by hand 'for the purpose of private study and review' would be a fair use. Users are now asserting that machine copying is merely a substitute for hand copying and is, therefore, a fair use. But this argument ignores the economic differences between the two types of copying. Copying by hand is extremely time consuming and costly, and is not an economic threat to authors. *Viewing reprography as though it were hand copying, however, overlooks the effect of the total number of machine copies made. Few people hand copy, but millions find machine copying economical and convenient. Allowing individual users to decide that their machine copying will not injure the author and will thus be a fair use fails to take into account the true economic effect when thousands of such individual decisions are aggregated.* [Emphasis supplied.]
. . .*Project, New Technology and the Law of Copyright: Reprography and Computers,* 15 U.C.L.A. L. Rev. 931, 951 (1968)." 487 F.2d 1345 (Ct. Cl. 1973), at 1368. *See* Appendix E, *infra.*

F. The Dual-Risk Approach Applied

Except for its awkward efforts to bring the exemption-like standards of educational photocopying into the ambit of fair use, section 107 of the new Copyright Act, embodying as it does, by design, at the very least no more in the way of fair use guidance than the courts have heretofore possessed,[95] cannot be expected to throw any further light on the sort of issues raised in fair use cases hitherto adjudicated. It remains for us to consider whether, with fair use more manageably defined, the dual-risk approach is any more helpful.

The general effect of the approach here, it will be observed from the analysis in section D, above, is not only to draw a sharper line between fair use and protected uses, but also to suggest some weighting of access and cost factors in accordance with the risks taken in the basic copyright-scheme design: (1) when what is involved are tensions *within* the copyright scheme, the balance is on the side of less protection of copyrighted works and is more solicitous of the society's interest in greater access; (2) when what is involved are tensions *between* competing copyright and noncopyright interests, the balance is on the side of greater protection of the copyrighted work and is more solicitous of the author's economic interests.

Also, with the priorities of the fair use factors thus set out, courts might find it easier to distinguish arguments *within* the copyright scheme from arguments *against* the copyright scheme. The dual-risk approach should thus have the effect, in particular cases, of raising earlier than they otherwise might have certain threshold questions the answers to which would argue for not reaching a fair use issue at all.

For example, the wording of section 107 of the new act is, unsurprisingly, of no help if applied to the facts of the two classic troublesome cases in the recent history of fair use adjudication—*Williams and Wilkins v. the United States*[96] and *Benny v. Loew's*,[97] pointing in opposite directions—on which the Supreme Court split 4-4. Under the dual-risk analysis proposed here, both cases would have very likely gone the other way.

In the *Williams and Wilkins* case, the Court of Claims held that

---

[95] "Section 107 is intended to restate the present judicial doctrine of fair use, not to change, narrow, or enlarge it in any way." *See* text at n. 51, *supra.*

[96] 487 F.2d 1345 (Ct. Cl. 1973), *aff'd by an equally divided court,* 420 U.S. 376 (1975).

[97] 239 F.2d 532 (9th Cir. 1956), *aff'd by an equally divided court,* 356 U.S. 43 (1958), *rehearing denied* 356 U.S. 934.

government agencies could systematically reproduce whole articles from various specialized medical journals and provide them free on request to the very scientists, libraries, and business firms for which they were written. Now, Congress has modified the effect of that decision in the new statute, by seeing the issue as an exemption question—a matter of reallocation of costs—and has accordingly dealt with it expressly in a sharply focused section of the statute.[98] But since neither the Court of Claims, which split 4-3, nor the Supreme Court could find in what they saw as fair use doctrine a clear bar to this sort of copying, it is hardly likely that the fair use section of the new statute, which supposedly changes nothing, would in a new case posing issues similarly fundamental to the copyright scheme guide courts to a position closer to the legislative view—if it *is* the legislative view—of fair use implied by the limitations in the photocopying exemption. The statute as it stands, as pointed out earlier, straddles the issue between sections 107 and 108 and retains the ambiguity, despite the straining efforts of the commentary in the committee reports to dispel it.[99] The random order of the "four factors" that permitted the Court of Claims in *Williams and Wilkins* to take its orientation almost wholly from the "needs" of the second user[100] would continue under section

----

[98] § 108(g). *See* Appendix A, *infra.*

[99] The doctrinal problem is no clearer even after the congressional committees expressly address it. Following is the comment in the Senate Committee Report on section 107 (the House version inserts the part in brackets):
  "The Register of Copyrights has recommended that the committee report describe the relationship between this section and the provisions of section 108 relating to reproduction by libraries and archives. The doctrine of fair use applies to library photocopying, and nothing contained in section 108 ['in any way affects the right of fair use.' No provision of section 108.] is intended to take away any rights existing under the fair use doctrine. To the contrary, section 108 authorizes certain photocopying practices which may not qualify as a fair use. The criteria of fair use are necessarily set forth in general terms. In the application of the criteria of fair use to specific photocopying practices of libraries, it is the intent of this legislation to provide an appropriate balancing of the rights of creators, and the needs of users." *See, e.g.,* Senate Committee Report at 67; House Committee Report at 74.

[100] This view, which reverses the presumption of the "basic risk" taken in the copyright scheme, leads Professor Nimmer to say that the *Williams and Wilkins* case "carried to its logical extreme would undercut the entire law of copyright." NIMMER ON COPYRIGHT, § 656.3.
  The Senate Committee Report on the *Williams and Wilkins* case in the discussion of section 108 dealing with library photocopying left the fair use—exempted use line in the same state of doctrinal and factual confusion in which the court in *Williams and Wilkins* left it: "It is still uncertain how far a library may go under the Copyright Act of 1909 in supplying a

107, while the dual-risk approach proposed here would begin with the appropriate expectations of the author of the first work.

On the other hand, *Benny v. Loew's,* in which infringement was found when the motion-picture *Gaslight* was burlesqued in a television sketch that took much of its dialogue and, of course, its plot, from the movie, would probably fall within the area of fair use under both the "necessity" and the "appropriately expected economic reward" tests of our proposed definition. There could be no doubting the "necessity" of using the actual words and story of the first work in a "take off" of this sort, unless the burlesque form were to be barred altogether;[101] and it is hardly likely that the "economic reward" factor could surmount both the "expected" and the "appropriate" tests so as to allow the author of the first work to control access. In other words, the balancing of the author's incentive and society's need for access that is inherent in the constitutional grant might have moved a judge, in applying the "four factors," to give greater priority to need for access and less to the mere quantity of the use.[102]

The effect of the dual-risk approach is sharply seen also in a consideration of the two recent and important cases of *Time, Inc. v. Bernard Beis Associates*[103] and *Rosemont Enterprises, Inc. v. Random House,*

---

photocopy of copyrighted material in its collection. The. . .case. . .failed to significantly illuminate the application of the fair use doctrine to library photocopying practices. Indeed, the opinion of the Court of Claims said the Court was engaged in a 'holding operation' in the interim period before Congress enacted its preferred solution. While the several opinions in the *Wilkins* case have given the Congress little guidance as to the current state of the law on fair use, these opinions provide additional support for the balanced resolution of the photocopying issue adopted. . .in section 108 of this legislation." Senate Committee Report at 71. The House Committee Report commentary on section 108 eliminates this passage and does not mention *Williams and Wilkins* at all. Neither does the Conference Report.

In short, neither the Court nor Congress is a help to the other in fashioning a more accurate general approach to fair use, however much the definition of the library copying exemption itself is clarified.

[101] Section 115(a) introduces a problem with respect to musical parodies. Section 115 grants compulsory licenses for the making of phonograph records to any manufacturer once the copyright owner has licensed and allowed to be distributed any recording of the work, but what the record maker can do with the music is limited to the requirements of the usual arrangement of a piece of music. It cannot be "perverted, distorted, or travestied," in the language of the committee reports. This would appear flatly to bar any parody of a piece of music without the copyright owner's permission. Would the principle apply equally to a *performance* of the music?

[102] The court in *Benny* seemed to rest its decision basically on the "substantiality" factor. 239 F.2d 532 (9th Cir. 1956).

[103] 293 F. Supp. 130 (S.D. N.Y. 1968).

Inc.,[104] in the first of which the court found infringement but fair use on a First Amendment theory, and in the second, fair use on the theory of right of access to factual material. The basic reasoning, if not the result, in both cases would be somewhat different under a dual-risk approach.

There would be no need, for example, to pose the sort of adversary relationship between fair use and First Amendment interests that Professor Nimmer suggests[105] in his discussion of *Bernard Geis,* where the court found that the use in a book of some frames from the Zapruder film of President Kennedy's assassination was fair use though infringing. The fair use test of "appropriately expected economic reward," modified by the necessity of access test, would encompass reasonable free-speech considerations without the unnecessary polarization of economic interests suggested by the court or Professor Nimmer in *Bernard Geis.*

As for *Rosemont,* Professor Nimmer is surely right in arguing[106] that the sort of copying involved could be justified neither on First Amendment grounds nor on fair use grounds. But the fair use issue was confused in *Rosemont* by the fact that the action was for an injunction by a Howard Hughes subsidiary that had been formed for the sole purpose of preventing publication of a book on Hughes by buying up copyrighted and published source material, namely a series of articles on Hughes in *Look* magazine. And the equitable nature of the case, triggering First Amendment and unclean hands factors in the judgment, also led the court to avoid facing the fair use issue squarely. The court did not decide whether the degree of copying was fair use, only that if it were not, both the determination of that issue and the remedy, if any, would have to be at law. The fair use issue would have been more clearly drawn if the court had simply found infringement, since the unacknowledged copying of exact parts of a copyrighted work was both clear and *unnecessary,* while denying the injunction and leaving the remedy (which was bound to be minimal) at law. The dual-risk approach, combining notions of appropriate expectations of payment with appropriate expectations of access, might

---

[104] 366 F.2d 303 (2d Cir. 1966).

[105] "The result in the *Bernard Geis* case can be defended, if at all, not on the ground of fair use, but rather because of the previously described free speech elements inherent in the film." Nimmer, *Does Copyright Abridge the First Amendment Guarantees of Free Speech and Press?* 17 UCLA L. REV. 1180, 1201 (1970).

[106] *Id.* at 1203.

more likely suggest such an outcome than an analysis in which the various factors are seen as both discrete and opposed.[107] On this analysis, then, it will be seen that because of the verbatim copying in *Rosemont,* the access question was perhaps a pseudo-fair-use issue, for all of the facts in the *Look* articles were available to a second user; and if that issue had not been complicated by the direct copying, there would have been even less of a threat to first amendment interests than the court found.

In certain other areas the greater precision of the proposed fair use definition would help courts to distinguish fair use issues from pseudo-fair use issues and thus to avoid jurisprudential confusion. Take, for example, the case of *Berlin v. E.C. Publications,*[108] on whose dicta *Rosemont* heavily relies. In *Berlin,* the copyright owners of such songs as "The Last Time I saw Paris" and "A Pretty Girl is Like a Melody" sued *Mad* magazine for infringement in publishing parodies of the lyrics, such as "The First Time I saw Maris" and "Louella Schwartz Describes her Malady," with wholly different words and themes and without the music. The court's opinion is full of good sense, and its eloquent dicta have been relied upon by courts in many subsequent fair use cases besides *Rosemont.* But *Berlin* under our approach would be seen, by any but the most absent-minded conceptualization, to be not a fair use case at all. Perhaps, indeed, not even a copyright case, for there is on the face of things no copying of either word or theme. Indeed, a good parody will in fact be a wholly new work, for which the parodied work is merely a point of departure. The *idea* of a parody cannot be copyrighted, whether nascent in the

---

[107] Professor Paul Goldstein, in *Copyright and the First Amendment,* 70 Col. L. Rev. 983 (1970), is concerned about a threat to First Amendment interests from both the "statutory monopoly" and what he terms the "enterprise monopoly" of publishing under copyright. After noting that "copyright's restraint upon what may be said and heard in public has not been noticed by the Supreme Court," at 984, and that only scant attention has been given to the matter by lower courts and scholars, he finds in the facts of the *Rosemont* case reasons for alertness to the danger of conflict between copyright and free speech. It is arguable, however, that since First Amendment interests have to do with the "appropriateness" of "expectations" with respect to access, they are at least as protected by fair use doctrine in a copyright context as in any other context in which competing interests threaten free speech. Statutory copyright under the Act of 1909, for example, in requiring publication, insured a significant degree of access and might therefore be seen as having left a very narrow First Amendment area subject to possible encroachment.

[108] 329 F.2d 541 (2d Cir. 1964).

original or express in a particular parody: anyone else can of course parody the work as well.

It is hard to see why the case was not briskly disposed of by the Second Circuit Court, but if we ignore the suspicion that the court did not want to deprive itself of more fun than it had had in a long time, the answer has perhaps partly to do with the fuzzy history of satire in the courts, particularly with respect to parody and burlesque.[109] Under dual-risk analysis, the court might have taken the opportunity to make some necessary and more precise distinctions between the essential elements of parody (more strictly, a new work in imitation of another's style) and the essential elements of burlesque (more strictly, an exaggerated and grotesque performance of another's work), rather than rely on such aids as dictionaries, which more often than not systematically obscure distinctions. Clarity is particularly required if court decisions are to hinge on such distinctions. In cases such as this, a court working under a statute embodying more expressly the relationship of the elements in the notion of fair use might more likely see that neither society's expectations with respect to access to the substance of the copyrighted work nor the author's appropriate expectation of economic reward was actually at stake, and thus be led to focus on the fact that there was simply no copyright issue at all.

The *Berlin* court took particular pains to reject ("For, as a general proposition, we believe that parody and satire are deserving of substantial freedom")[110] one of the basic findings of the *Benny* court: "No federal court . . . has supposed that there was a doctrine of fair use applicable to copying the substance of a dramatic work, and presenting it, with few variations, as a burlesque."[111] But though the *Berlin* dictum had nothing to do with the facts before it, it does have to do with the relationship between society's need for access and the author's expectation of reward. A dual-risk analysis would make it easier for courts faced with parody and burlesque facts to avoid both the literalness and the rigidity of the "exclusive rights" language and to see a different relationship between the sort of risks authors take with re-

---

[109] *See* NIMMER ON COPYRIGHT § 145, 648-51.

[110] 329 F.2d 541, 545 (2d Cir. 1964).

[111] 239 F.2d 532, 536 (9th Cir. 1956). The district court judge who decided *Benny* seems to have taken a less constricting view in a subsequent case, Columbia Pictures Corp. v. NBC, 137 F. Supp. 348 (S.D. Cal. 1955), but the Ninth Circuit adheres to *Benny*, holding that notwithstanding *Rosemont* and *Bernard Geis,* the public interest "is simply not relevant in determining fair use under the Ninth Circuit [parody] rule." Walt Disney Productions v. Air Pirates, 345 F. Supp. 108 (N.D. Cal. 1972).

Professor Nimmer argues that though the "function" of the second use

spect to access and the sort of promise society makes with respect to money.

Another example of the possible effect of the dual-risk approach on the threshold question is the case of *Henry Holt and Company v. Liggett and Myers Tobacco Company*,[112] cited by the *Rosemont* court. The *Holt* court held that the company's use in an advertisement of three lines from a doctor's medical treatise did not constitute fair use because the use was for a purely commercial purpose rather than to advance knowledge. But, again, the application of the two tests of author's expectations and society's need for access show this, as has been suggested,[113] as not a copyright case, but a privacy case. Society has need of unrestricted access to the knowledge, which should not be barred, but no unrestricted right to the use of the author's name.[114] Both interests could be accommodated outside the copyright scheme by permitting the advertisement to quote the knowledge, while attributing it to "an authoritative treatise" or the like.

The *Benny* case leads us into another case that it would be profitable to examine in the light of our dual-risk approach to these matters. The court in *Benny* relies on *Leon v. Pacific Telephone and Telegraph Co.*,[115] in which the telephone company brought suit against a small company and its owners who had published a phone directory of sorts, a "numerical telephone directory." Defendants had taken some of the information contained in the telephone company's alphabetical directory and rearranged it according to exchanges or prefixes, listing the numbers under the exchanges in numerical order, followed simply by the name of the person with that number. The court held that this was an infringement:

> "Counsel have not disclosed a single authority, nor have we been able to find one, which lends any support to the proposi-

---

can excuse copying, as in *Columbia Pictures,* "where there is. . .almost complete identity of content, then the functional distinction. . .becomes illusory," though the example conjured up ("incorporating substantially all of a short story in what purports to be a literary criticism of the story") seems much like a straw man, and the consequences he sees ("those who saw the Jack Benny burlesque. . .became rather fully familiar with the serious underlying story so that they might not thereafter wish to see the plaintiff's original work") unpersuasive. NIMMER ON COPYRIGHT, § 145, at 651.

[112] 23 F. Supp. 302 (E.D. Pa. 1938).

[113] B. KAPLAN, AN UNHURRIED VIEW OF COPYRIGHT 68 (1966).

[114] Goldstein, *Federal System Ordering of the Copyright Interest,* 69 COL. L. REV. 49, 91 (1969).

[115] 91 F.2d 484 (9th Cir. 1938).

tion that wholesale copying and publication of copyrighted material can ever be fair use."[116]

Two aspects of the same issue—society's expectations with respect to access—are raised, one having to do with the notion that for different kinds of works different thresholds of access are involved, and the other with the question of the extent of access within particular categories of copyrighted works. Some precision is foregone when these differences are so far lost sight of as to lead a court dealing with one kind of work to rely so heavily on cases where the nature of the first copyrighted work, which ought to be of great, if not controlling, importance, is intrinsically different from the second work. Neither *Leon* nor the only other case on which the *Benny* court heavily relies, *Chautauqua School of Nursing v. National School of Nursing*,[117] has to do with a work of the imagination, which invokes different expectations of society and author alike.

But another aspect of the issue raises perhaps even more interesting questions of fundamental fair use rationale. After the sentence quoted above in *Benny* ("Counsel have not disclosed . . .") the court continued: "The fact that plaintiff has not chosen to arrange its material in the inverted form used by appellant is no determinant of fair use,"[118] and goes on to cite precedent in England and the United States for the "settled" policy that the copyright holder cannot be prevented from barring a use of his work the owner has elected not to exploit.

The court in *Benny* assumes that this disposes of the question before it. But once the question posed is seen to be one of free access in the public interest, a more complicated copyright-scheme question is raised. Is the public-interest "access test" in the failure, say, of an author of a novel not to exploit his motion-picture rights to be the same as in the barring of the public to an appropriate and perhaps necessary use of a published work of information? For consider: What *Leon* did was to make an index to the telephone directory. Are indexes not furnished by an author to be barred? What interests of the author and of society are respectively at stake? Can the telephone company be said to have been deprived of an appropriately expected economic reward from the publication of an index they had no intention of publishing themselves? And can it be said that the telephone company

---

[116] *Id.* at 486 (9th Cir. 1938).
[117] 238 Fed. 151 (2d Cir. 1916), cited by the *Benny* court, 239 F.2d 532 (9th Cir. 1956).
[118] 91 F.2d 484, 486 (9th Cir. 1938).

was deprived of expected income from a work that it distributes free of charge as part of its service? If society finds that its need for access to the published material is important, is its choice between either requiring each person to make his own index or making do with none?

However these questions might be answered, the often-remarked upon different weight to be given the factors involved in creative works and informational works sharpens the fundamental question about the expectations of author and society.[119] The dual-risk approach to these questions emphasizes the chance the author takes, in accordance with the category of the work he produces, that free access will be required by society. What is suggested by the initial priority given to the "nature of the copyrighted work"—a priority with a governing function—is that the first author, in making public a work in one or another of these genres, will in law be considered to have recognized the different degree of control of access required in the public interest, and this will accordingly color the court's view of what constitutes "an appropriately expected economic reward" when a subsequent use of the work is considered.[120] It is a measure of the effect of this approach that its application to both *Benny* (with respect to creative work) and *Leon* (with respect to an informational work) could support findings of fair use.

We do not multiply examples. These are enough to indicate the general effect of this approach—a stricter view of an author's interests when he makes claims that disturb the copyright scheme, and a more relaxed view of the need for access when society makes claims consistent with the copyright scheme. It will be seen that even so there would remain difficult questions for the courts, on whose sense of equity, already substantially demonstrated in this area, there would of course continue to be reliance. Courts would still have to make judgments involving elastic terms like "expectations," "appropriate," "substantiality," "purpose," and "effect," and they would still have to engage in the kind of balancing involved in weighing how much the author's role in particular cases is congruent with the fundamental de-

---

[119] Points analogous to that developed in the text are made in NIMMER ON COPYRIGHT 171-72, and by J. Squires in *Copyright and Compilations in the Computer Era*, 24 BULL. CR. SOC. 18, 30 (1976). Though section 301 of the new Act does not limit rights under common law or state statutes, it is unclear how far state law could on a misappropriation theory control the use of compiled information seen to be permitted under fair use notions of copyright. *See* the discussion of section 301 in the Committee Reports. *See also* n. 121, *infra*.

[120] *See* Gorman, n. 87, *supra*.

sign of the copyright scheme.[121] But the elements of rigor put into the notion of fair use by the analysis, definition, and approach suggested here should do three things for courts: make more certain the threshold determination of the fair use line; provide an analytical scheme by which to determine whether or not there is a fair use issue at all; and establish some priorities in applying the factors relevant to the doctrine. Nor need courts wait for the ambiguities of section 107 to be expressly repaired: the analysis and approach suggested here are in no way incompatible with the existing law of fair use. But however unpersuasive this approach is in general, it is necessary for our particular purposes: we require the sharper line between protected uses and notions of fair use within the copyright scheme so as to deal with those exemptions from copyright control that Congress makes beyond that line—something to which we at last turn.

---

[121] Both the weakness of the fair-use definition in section 107 of the Copyright Act and the dangers of that weakness are exemplified by a late paragraph added in 1975 by the Senate Committee in its report and appearing, slightly changed, in the House Committee Report as follows:

"During the consideration of the revision bill in the 94th Congress it was proposed that independent newsletters, as distinguished from house organs and publicity or advertising publications, be given separate treatment. It is argued that newsletters are particularly vulnerable to mass photocopying, and that most newsletters have fairly modest circulations. Whether the copying of portions of a newsletter is an act of infringement or a fair use will necessarily turn on the facts of the individual case. However, as a general principle, it seems clear that the scope of the fair use doctrine should be considerably narrower in the case of newsletters than in that of either mass-circulation periodicals or scientific journals. The commercial nature of the user is a significant factor in such cases: Copying by a profit-making user of even a small portion of a newsletter may have a significant impact on the commercial market for the work."

House Committee Report at 73-74.

In short, the task is simply tossed to the courts, with on the one hand no bar to copying and on the other no more than an oblique warning (in the committee reports, not in the statute) to a user. The inclusion of the "appropriately expected economic reward" language in the statute itself would arguably give both a potential infringer and a court a clearer notion of the fair-use boundary. It is hard to avoid the thought that the committees would have done better not to say anything at all in the report about such information services.

III.

Exemptions from Copyright

A. The Rationale of Exemptions

With fair use we were concerned with drawing the line between protected uses and a use that the copyright scheme itself contemplates as not within the appropriately expected economic reward of the scheme. With exemptions we are concerned with what is on the protected side of that line—uses which are fully and properly within the copyright scheme but which should for reasons of public policy be declared exempt from the copyright control of the author.

In making this determination in a particular instance, Congress must deal simultaneously with two primary principles: the integrity of the copyright scheme, to which the public interest has on the whole been entrusted; and the strength of the call by other constitutional interests on modifying the internal dynamics of the copyright scheme. That is, just as initially a risk was taken in relying on copyright, so a risk will be taken when the scheme is significantly modified. What considerations ought Congress to take into account in deciding when to depart from the copyright scheme?

It is clear that if the fundamental reliance on the exclusive-rights mechanism of the design is to be relaxed it must be either for technical reasons, which would nevertheless leave the underlying incentive effect essentially intact; or for reasons of public policy deriving from the legislative view that the cost to society of exclusive author control is too great—that either the cost or the degree of control of access by the author, or both, is unacceptable. That is, unlike the primary fair use question of *whether* there is an appropriate cost at issue at all, the exempted-use question has to do with the *allocation* of acknowledged costs.

Having first come to a decision that there are appropriate copyright interests, whether of cost or of access, to be reallocated, Congress has two ways of dealing with an exception from the scheme: (1) either by altogether exempting certain uses from payment or permission, thereby concluding that no further reliance on the copyright scheme incentives is either needed or warranted; or (2) by substituting statutory for author controls of access and price (compulsory licensing), reaffirming the essential reliance on the copyright-scheme monetary incentives.

The conceptualization of an exemption as a reallocation of the access or economic costs attendant on a particular protected use explains why substantial technological change might be expected to put strains on the copyright scheme, for what technological changes do is to alter the *degree* in which the author can control access or price. It does not introduce a factor different in kind from what had to be considered when the initial copyright-scheme risk was taken. The central question in each case, then, is whether because of this change in degree of control, the initial reliance on the internal dynamics of the copyright scheme should be modified. The dramatic technological changes in recent years pose the question with particular force. The initial risk in the copyright scheme was taken when technological factors dictated that control of supply would be entirely in the hands of the author: he controlled the making of copies, the dissemination of his work to the public, and the cost to the public. When, however, the making of copies, such as with photocopying machines or tape recorders, is easily within the hands of the public itself; or when dissemination is possible by retransmission of signals freely available from the air or from recordings; or when these capabilities put cost factors within the control of the user—then the basic question is once again raised, at a high level of intensity.

But do these technological changes necessarily affect the validity of the underlying economic rationale? And if they do, does the basic copyright scheme any longer work in general? Or is it merely the case that the copyright scheme can no longer be relied upon in particular instances?

To get to these questions we must examine a little more carefully one of the principal characteristics of the property-like interest that the copyright scheme generates when it makes of a creation of the intellect a tangible commodity—namely, that it is easy to duplicate, to "take," in the sense that complete copying and transmission are possible in a way and to a degree that is not possible for any other commodity. The "taking" can occur in an instant; the "use" (by seeing or hearing) can be completely made at a distance of a thousand miles;

what is in other contexts merely the image of a thing is in copyright often the thing itself. There is simply a controlling difference between what must be done to duplicate a protected machine and a copyrighted work. What the second party does in duplicating a machine is essentially what the first one does: he builds it anew out of substantial and equivalent materials. There is a qualitative difference between that and a click of a copying device or the turning of a recording mechanism. It is this difference, of course, that makes both for the conception of copyright in the first place and for the continuing tensions within it; and it is precisely because of that characteristic that fundamental questions are raised anew by technological changes affecting the ease of duplicating copyrighted materials.

It becomes necessary as well to distinguish between two different meanings of "use"—use as embodied in a tangible medium of some sort, and use by the mind. The distinction is important in analytical terms, as it was in our discussion of the difference between a use by a second author of a first author's work and use involving a reproduction of an author's work for its own sake. A traditional formulation of the matter leads into the question.

The Register of Copyrights in his 1961 report to Congress says:

> "Copyright is . . . a form of property, but . . . intangible and incorporeal. The thing to which the property attaches—the author's intellectual work—is incapable of possession except as it is embodied in a tangible article such as a manuscript, book, record, or film."[122]

Now, there is another meaning to the notion of "possession" that is central to the question of technological impact on the copyright scheme: a work of the intellect can be "possessed" not only when it is embodied in a tangible medium, but also when it is perceived by another mind. A painting seen, a poem memorized, a piece of music heard has in that sense been "used" when it has simply been experienced by someone other than the author or the artist. That is one reason why a work of art is different from a machine, and that is of course why copyright law recognizes an author's interests in certain expressions of his work other than in tangible copies—in performances and displays, for example. But with respect to this kind of "use" it makes no difference whether the work has been read over someone else's shoulder, overheard by someone outside a concert hall, or observed on someone else's living room wall. The experience is the

---

[122] *Supra*, n. 33, at 3.

essential element in that sort of "use" of a work of the intellect. What a copyright scheme does is determine whether the author is to be paid when there is a transaction involving the means by which his work is transmitted or perceived. And it is when technological changes affect the economics of that transfer that the question of the design of the copyright scheme is raised.

Heretofore the copyright scheme has been able to accept the traditional commodity-economic model because of the fact that the author could on the whole control those means. The question now raised by technological changes is whether because he no longer can exert such control, the ordinary market-place economics can any longer be relied upon to reward him properly without undue costs to society. Again, the question is whether the difference in degree with respect to the characteristic of author control requires a change in a policy that relies on copyright economics. Is the risk to the public of high cost and of restraints on access so changed as to be unacceptable? And if that is in fact the case, so that those costs of access and price are no longer tolerable, what mechanism should we use to serve the public without weakening significantly the author's incentive? It is time to look at how Congress has answered these questions in the new Copyright Act.

B. The Exemptions in the New Copyright Act

Each exemption from copyright controls deals with both cost and access and is expressed either as complete freedom from payment and from control of access or as conditioned access, with statutory fees, in the form of a compulsory license. Of the following thirty-three exemptions culled from the Act,[123] four[124] have to do with compul-

---

[123] All but three are expressly set out in the statute, the others in the Senate and House Committee Reports in the commentary on fair use. Whatever the authority of legislative history in general, the necessity here for Congressional guidance on what is meant is so great, and expression of Congressional purpose so explicit as to warrant our treating several of the exemptions culled from the committee reports—numbers 8, 14, and 28 in the following catalogue in the text—as equal in legislative force to those of the statute itself. We have not included in this listing all of the exceptions expressly articulated in the committee comments on section 107 because they either redundantly express ordinary fair use notions (e.g., an individual may quote a derogatory statement in order to rebut it; or a congressional committee may quote a work in a legislative report), or deal with fair use issues at the margin (e.g., a student may make a single use of a piece of calligrapher's art).

As for the express statutory exceptions that we here uniformly call exemptions, the statute divides them into two categories, dealt with under

sory licensing. Twenty-seven of the 33 are grouped below in two categories, one covering copying and recording exemptions, the other performing and displaying exemptions; the remaining six, which raise complicating questions, are identified and dealt with separately.

1. Exemptions for Copying and Recording

The following exemptions apply to the making of copies or phonorecords.

Any library whose collections are available to the public or to "persons doing research in a specialized field"[125] can, if "without any purpose of direct or indirect commercial advantage,"[126] do the following without asking the copyright owner's permission and without paying fees:

(1) Make a facsimile copy or phonorecord of an unpublished work in its own collections for the collection of another such library, unless the library is expressly barred by an agreement with the copyright owner.[127]

(2) Make a facsimile copy or phonorecord for its own use of an out-of-print work it has owned and lost if it cannot buy a new one at a "fair price,"[128] irrespective of whether or not the copyright owner is willing to grant such a license to copy, with or without royalty.

(3) Make a facsimile copy or phonorecord of a damaged or deteriorating work to replace a work in its own collections if it cannot buy a new one at a "fair price."[129]

(4) Make for an individual patron,[130] for his or her private use,[131] a single copy[132] or phonorecord of a work that the library has first

---

the headings "limitations on exclusive rights" (sections 107-112, 118) and "scope of exclusive rights" (sections 113-117). Section 117, which has to do with computers, does not create an exemption from copyright, its intent being merely to preserve the status quo with respect to computer uses of copyrighted material and with the copyrightability of computer programs until Congress has acted on the recommendations of the National Commission on New Technological Uses of Copyrighted Works (CONTU). This book accordingly does not deal with aspects of copyright having to do with computers. See Appendix A, *infra.*

[124] Numbers 12, 23, 24, and 25, *infra.*
[125] § 108(a) (2).
[126] § 108(a) (1).
[127] § 108(b), 108 (f).
[128] § 108(c).
[129] § 108(c).
[130] § 108(e) (1).
[131] § 108(e) (1).
[132] § 108(a).

determined cannot be bought at a "fair price"[133]—except for a separate musical work, a pictorial, graphic, or sculptural work, or a motion picture or other audiovisual work other than an audiovisual work dealing with news[134]—so long as the copy of the work that is copied comes from the collections of its own or some similar library's[135] and so long as the copying is not "systematic" or "concerted."[136]

(5) Make a single copy or phonorecord at a time for private use of "no more than one article or other contribution" to a collection or periodical issue, or of a "small part" of any other work, whether or not the contribution or small part of the work is separately for sale by the copyright owner, and whether or not the library owns a copy of the work, so long as the work copied comes from some similar library, so long as the copying is not "systematic" or "concerted," and so long as the work is not a separate musical work, a pictorial, graphic, or sculptural work, or a motion picture or other audiovisual work other than an audiovisual work dealing with news.[137]

(6) Make and distribute to anyone a "limited number" of copies of an audiovisual news program,[138] which presumably could also be performed by the party to whom the copy is distributed.

In addition:

(7) Pictures of useful articles incorporating copyrighted pictorial, graphic, or sculptural works may be freely used for advertising, sales, or news purposes.[139]

(8) Any individual may make a single copy or phonorecord of a copyrighted work as a free service for a blind person.[140]

(9) A broadcaster or other transmitting organization may make a single copy or phonorecord of its own authorized and permitted broadcast of a copyrighted work (except for a motion picture or other audiovisual work) for its "own use" within six months, after which it may be kept solely for archival purposes.[141]

(10) A governmental body or other nonprofit organization may make and distribute to similar organizatons, for a period of seven years, without payment of royalty, thirty copies or phonorecords of a

---

[133] § 108(e).
[134] § 108(h).
[135] § 108(e).
[136] § 108(g) (2).
[137] § 108(d), 108(g) (2), 108(h).
[138] § 108(f) (3).
[139] § 113(c).
[140] House and Senate Committee Reports on § 107.
[141] § 112(a).

transmission of a performance of a nondramatic literary or musical work made as part of systematic instructional activities, provided that at the end of seven years all copies are destroyed save for one archival copy.[142]

(11) A governmental body or other nonprofit organization may make and distribute to a licensed transmitter, for a period of one year, copies or phonorecords of a transmission of a nondramatic musical work "of a religious nature," for the sole purpose of a single retransmission by each of the licensees, provided that at the end of the year all copies are destroyed except for one archival copy.[143]

(12) A governmental body or other nonprofit organization may make ten copies or phonorecords of a performance of a nondramatic literary work made for the blind or other handicapped persons for the sole purpose of a retransmission of the same sort.[144]

(13) Once the copyright owner of a nondramatic musical work has authorized the distribution of a phonorecord of the work to the public, any person whose "primary purpose" in making phonorecords is to distribute them to the public for private use may obtain a compulsory license, with statutory royalties, to make and distribute phonorecords of the work.[145]

(14) Copies of old motion-picture films subject to deterioration (mainly pre-1942 prints) may be made for archival preservation.[146]

2. Exemptions for Performances and Displays

The following exemptions apply primarily to performances and displays.[147]

(15) The owner of the physical embodiment of a work, whether of the original (as in a work of art) or of a copy, may without the consent of the copyright owner display it publicly in a gallery or display case, either directly or by means of a projector, "no more than one image at a time [which rules out, for example, motion pictures] to viewers present" at the site of the work.[148]

(16) Any nonprofit educational institution may freely perform or

---

[142] § 112(b), § 110(2).
[143] § 112(c).
[144] § 112(d).
[145] § 115.
[146] Senate and House Committee Reports on § 107.
[147] *See* in § 101 the definitions of "to display" and "to perform", as well as the definition of "to perform and display a work 'publicly.' "
[148] § 109(b), Senate and House Committee Reports on § 109.

display any work, provided it is done by "instructors or pupils in the course of face-to-face teaching activities . . . in a classroom."[149]

(17) Any nonprofit educational institution or governmental body may either directly or by transmission (rather than merely face-to-face), or by a temporarily made recording because of time-zone problems, perform a "nondramatic literary or musical work" or display any work if it is "a regular part of . . . systematic instructional activities . . . directly related and of material assistance to the teaching content" and done "primarily" for reception either in classrooms or anywhere else reachable by television if "aimed at [though not limited to] regularly enrolled students and conducted by recognized higher educational institutions" or at others who are either government employees at work or "who cannot be brought together in classrooms, such as preschool children, displaced workers, illiterates, and shut-ins."[150]

(18) A church may perform a dramatico-musical, musical, or non-dramatic literary work of "a religious nature," and may display any other work, provided that the performance or display is in the course of a service at a place of worship and that it is not transmitted to any dissimilar place.[151]

(19) Anyone, for a noncommercial purpose, may publicly perform (but not display and not transmit further to the public) a nondramatic literary or musical work, provided there is no admission charge and none of "the performers, promoters, or organizers" is paid a fee or other special compensation for the performance.[152]

(20) A single radio or television "apparatus of a kind commonly used in private homes" may be used to perform or display the transmission of any work to the public, provided no direct charge is made to see or hear the transmission and provided the radio or television signal is not retransmitted "beyond the place where the receiving apparatus is located."[153]

(21) A governmental body or nonprofit agricultural or horticultural organization may freely perform a nondramatic musical work at

---

[149] § 110(1).

[150] § 110(2), § 111(a) (2), § 112(b), House and Senate Committee Reports on §§ 107 and 110.

[151] § 110(3), House and Senate Committee Reports on § 110.

[152] § 110(4) (A).

[153] § 110(5). The permissible line is not all that clear, depending as it does on how a court might interpret the meaning of "small" and "relative" in the phrase "a small commercial establishment. . .with. . .loudspeakers grouped within a relatively narrow circumference from the set." *See* n. 175, *infra.*

an annual agricultural or horticultural fair conducted by such body or organization.[154]

(22) A retail record store may publicly play a recording of a non-dramatic musical work provided it does not transmit the sound beyond the "immediate area where the sale is occurring."[155]

(23) Noncommercial educational radio and television stations may freely perform a nondramatic literary work in broadcasts designed for the blind or deaf.[156]

(24) A cable system may have a compulsory license, with statutory fees and notice requirements, to retransmit (to "perform") certain publicly broadcast local or distant television or radio signals incorporating a copyrighted work.[157]

(25) The operator of a coin-operated phonorecord player (jukebox) may obtain a compulsory license, with statutory fee, to play records of a copyrighted nondramatic musical work in a public place, provided no admission fee is charged.[158]

(26) A public broadcasting "entity" has a compulsory license, with royalties determined by statute and by the Copyright Royalty Tribunal, to broadcast any published nondramatic musical, pictorial, graphic, or sculptural work, absent a voluntary agreement between the copyright owner and the public broadcasting entity.[159]

(27) A radio subcarrier may make for blind audiences a single performance of a dramatic work published at least ten years earlier.[160]

3. Other Copying and Performing Exceptions

In addition to these exemptions, which come wholly clear-cut from the text of the statute or from the legislative history as embodied in the committee reports, there are certain other uses of copyrighted material free of copyright controls that have the same outcome as the explicit exemptions listed above but because of certain characteristics need to be examined separately. There are three of these: (1) certain

---

[154] § 110(6).

[155] § 110(7).

[156] § 110(8).

[157] § 111(c), 111(d).

[158] § 116.

[159] § 118. The section also provides for an exemption from antitrust laws (an exemption not included in this catalogue of exemptions from *copyright*) so that "voluntary" uniform licensing agreements may be worked out for public broadcasting of nondramatic *literary* works.

[160] § 110(9).

exempted uses characterized by the statute as examples of fair use, (2) performing rights in sound recordings, and (3) certain exclusions having to do with secondary transmissions, particularly cable systems. Among them we identify six more exemptions.

a. *Fair Use "Exemptions."* As has been suggested in Chapter II, Congress has in the text and commentary of section 107 confused fair uses with exempted uses, partly because it has not in general terms defined the protected area of copyright and partly because it has used the committee reports on the fair use section to deal with educational exemptions for photocopying instead of making those exemptions in a separate section of the statute itself. Other educational exemptions from copyright are dealt with by Congress expressly and directly in the statute, rather than in the committee reports. For example, section 110 sets out precise rules on educational and school exemptions from copyright controls of performances and displays; section 111 deals with precision with secondary transmissions, including cable systems, for school and instructional use; section 112 deals expressly with school use of ephemeral recordings; and the section 108 exemptions for library copying would apply to educational libraries insofar as they meet the expansive institutional definitions of the section.

In short, only with respect to the photocopying and recording of copyrighted materials for school use does the statute obscure the issue. The reason is clear from, though not excused by, the legislative history: educators argued for a complete educational exemption in the hearings before the Congressional committees,[161] and when this was

---

[161] "Over the years, some of the educators have seemed to be arguing that, with respect to photocopying, they enjoy under the present law a 'not-for-profit' limitation co-extensive with that applicable to certain performances, and that somehow this 'right' is being taken away from them. This line of argument tended to produce a rather testy reaction, since plainly the only explicit 'not-for-profit' limitations on the copyright owner's exclusive rights under the present law are with respect to public performances of nondramatic literary and musical works. On the other hand, although the commercial or nonprofit character of a use is not necessarily conclusive with respect to fair use, in combination with other factors it can and should weigh heavily in fair use decisions. It would certainly be appropriate to emphasize this point in the legislative commentary dealing with fair use and educational photocopying." Second Supplementary Report of the Register of Copyrights on the General Revision of the U.S. Copyright Law: 1975 Revision Bill ch. II, pp. 27-28 (October-December 1975).

This "commercial or nonprofit" formulation, which seems to be without content, so recommended itself to the House Judiciary Committee that for the first time it incorporated the notion in the text of the statute itself. (*See* n. 164, *infra.*) Theretofore the "four factors" had with respect to the

denied, on the grounds that it would undermine the basic copyright scheme, they joined educational publishers in fashioning a fair use text that permitted the committee report to spell out a degree of exception from copyright controls. What it also did in the process was both to muddy the conceptualization of fair use and to introduce an element of obscurity about the fair use line in other contexts as well as in the school-use context. For it is arguable that if the exceptions were such as could not be derived by courts from the public policy shape of the fair use doctrine itself, then they must necessarily be exemptions. And that the outer standards could not in fact be determined by the courts is at once asserted by the commentary's suggestion that the boundaries can be enlarged or narrowed by the parties themselves. The committee reports' conclusory explanation for all this—that "a specific exemption freeing certain reproductions of copyrighted works for educational and scholarly purposes from copyright control is not justified"—is simply not persuasive.

To keep our bearings with respect to what exemptions the new copyright act decrees, then, we must ignore the Congressional conceptual imprecision and derive such further exemptions as we can from the combination of fair uses and exempted uses set out in the Committee Reports on section 107. Two of these have been included in the list above, the making of single copies for the blind (number 8)[162] and the copying of old motion-picture films in order to preserve them (number 14).[163] The third is the educational exemption, which we simply number (28) here, without trying to distill its meaning, except to say that it has to do with the classroom situation. Here we simply refer to the long and involved sections of the House and Senate Committee Reports[164] dealing with the matter. What it all might mean we take up below, in our discussion of public policy exceptions from copyright.[165]

b. *Performing Rights in Recordings.* However one might view the distinction between what is an *exclusion* from copyright (a work or use not within the scope of copyright at all) and an *exemption* (something covered by the copyright scheme but specially exempted from its exclusive controls), it is necessary to consider separately the question of

---

reprinting of a work been unencumbered, in either judicial or legislative history, by the conceptually unintelligible "nonprofit" element in a nonexemption context.

[162] *Supra,* at note 141.

[163] *Supra,* at note 147.

[164] House Committee Report at 66-72; Senate Committee Report at 61-67. *See also* n. 207, *infra.*

[165] *See* text at n. 207, *infra.*

statutory control of the right of performance of a copyrighted sound recording.[166] Since the right to *copy* and the right to *perform* are distinct rights, if a copyrighted musical work, for example, is recorded, there are four separate "rights" in question, two (copying and performing) having to do with the copyright in the musical work itself and two with the copyright in a particular recording of that work. Three of these rights are granted exclusively by the statute—the rights to copy and to perform possessed by the copyright owner of the musical work, and the right to copy possessed by the copyright owner of the recording. The fourth—the exclusive right to play (i.e., "perform") the record in public—is expressly denied the owner of the copyright in the recording.[167] That is, when a radio station plays a record of a copyrighted song, only the owner of copyright in the song receives royalty from the broadcast; the owner of the separate copyright in the recording does not. Also, if the recording is of a work in the public domain, no royalty for or control of performance is due the owner of the copyright in the recording.

Though this performing right in a recording has never been within the scope of the copyright scheme, and still is not, it nevertheless comes close enough, by virtue both of its legislative history and of the analogous performing right in the underlying piece of music, to the notion of an exemption to be identified here as such, and we accordingly number it (29) in our catalogue. The statute expressly requires the Register of Copyrights to submit a report to Congress on January 3, 1978, with recommendations about whether to give performing rights in records copyright protection.[168]

c. *Certain Secondary Transmissions and Other Problems.* Completion of a catalogue of copyright exceptions requires addressing ourselves to certain matters dealt with in section 111 of the Act called "secondary transmissions" (cable systems, for the most part) that with respect to the exemption-exclusion distinction fall more on the exclusion side of the line, rather than, as with performing rights in records, on the exemption side. This result derives from the definition of certain technical devices as merely "passive" relayers to private users of signals already covered by copyright controls. There are four such exclu-

---

[166] § 114.

[167] § 114(a): "The exclusive rights of the owner of copyright in a sound recording. . .do not include any right of performance. . .".

[168] § 114(d). The 1974 Senate Copyright Revision Bill, S. 1361, 93d Cong., 2d Sess., included such a performing right; it was deleted by the Senate Committee in its 1975 bill, S. 22. In 1975, separate House (H.R. 5345) and Senate (S. 111) bills for copyright protection of performances of sound recordings, with compulsory licensing features, died in Congress.

sions, to which we give the next four numbers in our list of exemptions: (30) a hotel, apartment house, or similar establishment may retransmit a local FCC-licensed radio or television signal to private rooms, provided no extra charge is made, without triggering new copyright controls;[169] (31) a common-carrier ("passive") cable system does not trigger copyright controls;[170] (32) a governmental or other nonprofit signal-boosting system, charging only at cost,[171] does not trigger copyright controls, nor (33) does a pay cable system that is required by the FCC to carry a coded pay television signal, thereby acting as a sort of common carrier.[172]

These are seen by Congress not as "performances" of copyrighted works, but in terms of copyright controls more like the expected sort of "uses" as usually go unremarked upon in a statute, such as the private reading of a book, the turning on of a private radio or television set, or the singing of a song over the telephone to a friend—that is, a use so "frictionless," so to speak, as not even to require fair use lubrication. But that has not always been the case with particular mechanisms, and the exclusions dealt with in section 111 are expressly addressed by Congress in the statute because their novelty at particular times in the technological development of electronic transmissions raised important and troublesome questions for the courts about whether these in fact constituted a "use" or "performance" such as to invoke copyright controls.

In *Fortnightly Corp. v. United Artists Television, Inc.,*[173] and *Teleprompter Corp. v. Columbia Broadcasting System, Inc.,*[174] cases having to do with the relaying of primary television signals by commercial CATV systems, the Supreme Court saw all such relaying activities as strictly technological, and therefore not infringing of copyright. Congress in section 111 accepts this "passive" view only insofar as there is no intervening commercial factor. (Exemption number 20 derives from this view.) Beyond nonprofit or common-carrier relaying, all commercial cable television is subject to the copyright statute, reversing the *Fortnightly* and *Teleprompter* holdings.[175]

---

[169] § 111(a) (1).
[170] § 111(a) (3).
[171] § 111(a) (4).
[172] § 111(b).
[173] 392 U.S. 390 (1968).
[174] 415 U.S. 394 (1974).
[175] Something of the same distinction is made with respect to the public use of a radio or television set. Congress drew a line in section 110(5) between the public playing of a single set in a particular place, which is permissible (exemption 19, above), and the further transmission of a signal to the pub-

But that does not end the matter so far as our identification of exceptions from copyright controls is concerned, for though cable systems have theoretically no technological limits with respect to the number of channels they might bring to a private subscriber, they impinge on the market for broadcast signals, which do have such limitations. Accordingly, the responsibility of the Federal Communications Commission to insure the viability of licensed broadcasting stations is seen to require it to have control over local transmissions, whether by wireless or cable.[176]

The cable-transmission problem is dealt with in the statute by the making of three distinctions and by the use of three corresponding

---

lic beyond the place where the receiving set is located, which *would* trigger copyright controls. In so doing, Congress upheld the outcome on the particular facts of the case in Twentieth Century Music Corp. v. Aiken, 422 U.S. 151, 95 S. Ct. 2040 (June 17, 1975), which held that the public relaying of a radio signal by a small restaurant by means of a few near-by extension speakers in the same room was the same order of phenomenon as the retransmissions of signals in *Fortnightly* and *Teleprompter*. More important, Congress returned to what had more or less been seen before *Aiken* to be the law since the Supreme Court's decision in Buck v. Jewell-LaSalle Realty Co., 283 U.S. 191 (1931)—namely, that a substantial commercial pick-up of a broadcast signal *does* trigger copyright controls. The line is not all that simple, however, as the commentary in the House Committee Report on 110(5) demonstrates:

"Under the particular fact situation in the *Aiken* case, assuming a small commercial establishment and the use of a home receiver with four ordinary loudspeakers grouped within a relatively narrow circumference from the set, it is intended that the performances would be exempt under clause (5). However, the Committee considers this fact situation to represent the outer limit of the exemption, and believes that the line should be drawn at that point. Thus, the clause would exempt small commercial establishments whose proprietors merely bring onto their premises standard radio or television equipment and turn it on for their customers' enjoyment, but it would impose liability where the proprietor has a commercial 'sound system' installed or converts a standard home receiving apparatus (by augmenting it with sophisticated or extensive amplification equipment) into the equivalent of a commercial sound system. Factors to consider in particular cases would include the size, physical arrangement, and noise level of the areas within the establishment where the transmissions are made audible or visible, and the extent to which the receiving apparatus is altered or augmented for the purpose of improving the aural or visual quality of the performance for individual members of the public using those areas."

[176] The Supreme Court in United States v. Southwestern Cable Co., 392 U.S. 157 (1968) upheld the F.C.C.'s authority to regulate cable systems, which the agency had asserted under an order promulgating rules issued on May 8, 1966.

approaches in the Copyright Act: (1) the categorization of nonprofit
"booster" systems or "common-carrier" facilities as outside copyright
altogether;[177] (2) the blanket inclusion of all commercial cable systems
in the copyright scheme by means of compulsory licensing, with royalties based on overall subscription income of the cable systems;[178] and
(3) the allocation of market rights through a dual compulsory licensing system by means of (*a*) statutorily defined markets—local and distant radio and TV signals [exemption 23, above], and (*b*) FCC-
determined allocations of other market rights and of certain signal retransmissions.[179]

The section of the statute that deals with the FCC's role simply
permits a compulsory license under certain conditions "where the carriage of the signals comprising the secondary transmission is permissible under the rules, regulations, or authorizations of the Federal
Communications Commission."[180] This may appear to be unduly expansive and open-ended, but it does not in fact either make or enlarge an exemption from *copyright.* That exemption has already been
made in the statute by the compulsory licensing language for retransmissions that includes all cable systems in general: what the FCC
provision does is simply to give to the agency the job of regulating the
already exempted use of copyrighted material among competing systems.

So much for the exemptions and exclusions express or implied in
the statute. There remains to be examined an anomaly in the act, a
peculiar pseudo-exemption the characteristics of which lead us naturally into the considerations of rationale dealt with in the next section
of this chapter of this book. Section 110, headed "Limitations on exclusive rights: Exemptions of certain performances and displays," covers
the complete performance exemptions from permission and royalty of
certain works for such uses as instruction in schools, a part of a
church service, and the other exemptions included in numbers 16
through 23, above. But subsection (4) of section 110 reads as follows:

> [Notwithstanding the provisions of section 106, the follow-
> ing are not infringement of copyright]:
> (4) performance of a nondramatic literary or musical work
> otherwise than in a transmission to the public without any pur-
> pose of direct or indirect commercial advantage and without

---

[177] § 111(a) (1), (2), and (4), 111(b).
[178] § 111(b), (c), (d), (e).
[179] § 111(c), (d).
[180] § 111(c).

payment of any fee or other compensation for the performance to any of its performers, promoters, or organizers, if:

(A) there is no direct or indirect admission charge, or

(B) the proceeds, after deducting the reasonable costs of producing the performance, are used exclusively for educational, religious, or charitable purposes and not for private financial gain, *except where the copyright owner has served notice of his objections to the performance under the following conditions:*

(i) the notice shall be in writing and signed by the copyright owner or his duly authorized agent; and

(ii) the notice shall be served on the person responsible for the performance at least seven days before the date of the performance, *and shall state the reasons for his objections;* and

(iii) the notice shall comply, in form, content, and manner of service, with requirements that the Register of Copyrights shall prescribe by regulation; . . .

[Emphasis added.]

What are we to make of the curious non-exemption described in the "except" clause of subparagraph (B)? Though the statute does not in terms require notice *to* the copyright owner about the forthcoming performance, presumably the Register of Copyrights will make a ruling not only requiring such notice but also providing for enough time to enable the copyright owner to respond in writing in accordance with the notice requirements of the statute. With that implementation, since permission for a paid-admission use, for whatever purpose, must be obtained from the copyright owner, *there would appear to be no exemption at all.*

But the subsection is not without two effects, one ambiguous, the other clear. The ambiguity lies in the possibility of the statute's being interpreted to mean that if the copyright owner grants permission he cannot charge a fee; and as we shall see, there is wording in the House Committee Report that suggests such a possibility. Though that interpretation raises questions on which it is fruitless to speculate, it is the only one that permits an "exemption" meaning to the provision at all. What is clear, however, and disturbing, is that a condition of the copyright owner's refusal to grant permission is that he must under subsection B(ii) publicly explain why—surely an unconstitutional invasion of rights of freedom of speech and of privacy. If he does not, and "in writing," may the sponsors proceed to use his work without permission? When a court gets round to considering the issue, it may or may not be helped by the legislative history of the subsection as revealed in the Committee reports accompanying the various bills.

The "except" clause, word for word as it appears in the Act, was added by the House in the 1967 bill[181] and explained in the Committee Report in language much like that in the 1976 House Committee Report, which states that "the copyright owner is given the opportunity to decide whether and under what conditions the copyrighted work should be performed; otherwise owners could be compelled to make involuntary donations [may owners *not* charge a fee?] to the fund-raising activities of causes to which they are opposed. The subclause would thus permit copyright owners to prevent public performances of their works . . ."[182] But not, it is made clear, without explaining themselves.

Congress has absently walked up to a question having to do with a substantive adjustment between copyright-scheme interests and other constitutional interests—something that requires a *reallocation* of costs—belatedly seen that something was wrong with its approach, and in the "except" clause hastily retreated, leaving the fundamental question only dimly seen. Innocuous as it nevertheless is here, this example of conceptual confusion might be an uneasy sign, if we were looking for one, that however successful Congress might be in avoiding most conceptual pitfalls, it might occasionally have trouble finding a coherent compromise among competing interests.

In any case, this puzzling provision, awkwardly straddling several matters of public policy, poses the question ultimately to be faced with respect to the new Copyright Act as a whole: In what way, if any, do the changes in control of exclusive rights embodied in the new Copyright Act affect the fundamental design of the copyright scheme? It is to a consideration of that question that we now turn.

C. Analysis of Exceptions from "Exclusive Rights" Controls in the New Copyright Act

1. Classification of Exceptions: Technological or Public Policy

Of the thirty-three exemptions and other exceptions identified in section B, above, some of which have to do with several forms of use,

---

[181] H.R. 2512, 90th Cong., 1st Sess. (1967).

[182] House Committee Report on the 1976 bill, S. 22. The Senate Committee Report accompanying the 1974 bill, working with the House Committee's 1967 text, omitted the rationale but stated that ". . .if there is an admission charge the copyright owner may prevent a public performance of his work. . ." Report to Accompany S. 1361, at 129. But the final 1975 Senate Committee Report, on S. 22, deleted even *that* comment and says nothing at all about the provision. This sequence might suggest that the Senate

twelve apply to the reproduction of printed matter or other visual material, twelve to the recording of aural material, and nineteen to the performing, displaying, and transmitting of a work, either directly or from a captured image or signal. Four of these (numbers 13,[183] 18,[184] 19,[185] and 25[186]), more or less in this form, were express exemptions in the 1909 act, and a fifth (number 29,[187] on performances of recordings), an exclusion (i.e., beyond the scope of copyright protection altogether), was likewise not within the scope of the old act. Of the rest, some are wholly new, some derive from court decisions that Congress has confirmed, modified, or reversed, and some simply recognize past practices that while perhaps technically infringing have never actually been challenged in court.

At the very least, they represent the codification of a significant body of adjustments and modifications of the copyright scheme. To approach the larger question—"Do they in the aggregate represent a fundamental change in public policy with respect to the basic scheme of copyright?"—we need first to distinguish changes that essentially constitute the necessary accommodation of new technology from those that primarily reflect new substantive public policy determinations. We accordingly set out the exemptions and exclusions into these two categories in Tables I and II, drawing the line rather strictly between access (technological) accommodations and cost allocation (economic) provisions.

Table I. Classification of Exemptions and Exclusions from Copyright for Technological Reasons: Access Accommodation

A. *Passive Exclusions*
   1. *Performing*
      (a) Secondary transmissions by nonprofit boosters (32)
      (b) Secondary transmissions by common carriers (31, 33)
      (c) Secondary transmissions by noncommercial relays of
         broadcasts to private lodgings (30)
      (d) Public performances of a broadcast by ordinary
         receiving instruments (21)

Committee saw something embarrassing in the provision, but not enough to disturb it.

[183] *See* text at n. 146, *supra.*
[184] *See* text at n. 151, *supra.*
[185] *See* text at n. 152, *supra.*
[186] *See* text at n. 158, *supra.*
[187] *See* text at n. 166, *supra.*

2. *Reproducing*
    (a) Ephemeral recordings by broadcasters (9, 11, 17)

B. *Compulsory Licenses*
  1. *Performing*
    (a) Commercial secondary transmissions (cable systems) (24)
    (b) Jukebox performances of musical works (25)
    (c) Public broadcasting entities (26)
  2. *Reproducing*
    (a) Phonorecords of musical works (13)

Table II. Classification of Exemptions and Exclusions from Copyright for Public Policy Reasons: Cost Allocation

A. *Peripheral Economic Adjustments: Essentially not part of the usual commodity-market mechanism*
  1. *Performing or Displaying*
    (a) Public display by owner of tangible work (15)
    (b) Record playing by retail record stores (22)
  2. *Reproducing*
    (a) Use of picture of useful article in advertising or news report (7)
    (b) Library preservation of old motion-picture films (14)
    (c) Library preservation of deteriorating or damaged out-of-print works (3)
    (d) Library reproduction of an unpublished work in its own collection for another library (1)

B. *Integral Economic Adjustments: Essentially accommodatable by the normal commodity-market mechanism*
  1. *Performing or Displaying*
    (a) Educational use (10, 16, 17)
    (b) Noncommercial use (19)
    (c) Religious use (11, 18)
    (d) Agricultural-fair use (21)
    (e) Use for blind, deaf, and other handicapped (12, 23, 27)
    (f) Public and private use of sound recordings (29)
    (g) Audio-visual news broadcasts (6)
  2. *Reproducing*
    (a) Educational use (28)
    (b) Use for the blind (8)
    (c) Audio-visual news broadcasts (6)
    (d) Library copying (2, 4, 5)

The classification into two tables has been made without regard to the degree of economic adjustment, though for purposes of determining the general shape of the copyright scheme under the Act, the economic weight of particular provisions must in due course be considered. Nevertheless, some of the cases set out in Table I as "for technological reasons" do involve certain aspects of public policy that might be seen as at least marginally economic, and some of the cases listed as "for public policy reasons" in Table II might be seen as so minimally economic as to fall more appropriately on the technological side. Two uses, numbers 11 and 17 in our exemptions catalogue, having to do primarily with exemptions for the performance of non-dramatic music in schools and churches, and therefore primarily substantive in nature, are also involved in a technologically required accommodation on non-policy grounds, and therefore appear in both tables.

But though there is one anomaly[188] among the twelve items in Table I, the intent in the classification there is to include those that are essentially undisturbing of the primary balancing in the copyright scheme and that would be required by technological factors alone even if there were no substantive factor at all. But the disorienting force of swift technological change typically outruns the law, and none of the eleven "technological" terms in Table I has been unembattled. The classification accordingly needs some explanation.

2. Technological Exceptions.

The "passive" exclusions in Table I, some of them after years of argument during congressional hearings, have in the end simply been so defined by Congress. Such of them as might arguably have been defined differently, like those having to do with secondary transmissions (essentially cable television systems), are defined in terms categorizing them as undisturbing of copyright-scheme economics. That conceptualization saves the Congressional solution from what might otherwise be seen as a fatal circularity. That is, so long as the governing structural elements in the copyright scheme are clear, the internal economic workings can accommodate them. Once, therefore, having defined these factors as nondisturbing of copyright-scheme economics, Congress could deal with the cable television problem as essentially an access question rather than one of cost reallocation. In

---

[188] The compulsory licensing for public broadcasting entities, § 118, exemption 26, *supra. See* n. 190, *infra.*

that context, then, the secondary transmission cases most reasonably fall on the purely "technological" side of the line.

The part of the classification that might raise the strongest questions is that having to do with compulsory licenses, which are sometimes seen as in fact threatening the entire copyright scheme.[189] But except for the exemption for public broadcasting entities, which unnecessarily establishes a government-operated scheme parallel to a market licensing mechanism already in operation,[190] the other three compulsory license schemes (for making records of a piece of copyrighted music, for jukebox playing of records of copyrighted music, and for commercial cable television use of copyrighted works), though with features arguably unnecessary in a statute, leave the copyright scheme essentially intact.[191] These compulsory licenses are

---

[189] "Throughout the whole range of national and international copyright regimes since 1950, a single concept insistently recurs: it is usually called compulsory licensing. . .[T]he author loses the right to control the use of his work, and cannot grant anyone an exclusive license for a certain specified purpose. . .[He] becomes a unit in a large collective system under which blanket royalties are received and distributed. The government is involved in operating the system, and the individuality of both authors and works tends to be lost. . .[A] new and very significant institution, a royalty tribunal. . .would create a government-associated body empowered to make decisions with respect to the practical running of the compulsory licensing system. . .We have reached the point where any new rights under the copyright law apparently cannot be exclusive rights. If a new technological development makes new forms of exploitation possible, compulsory licensing seems to offer the only solution. . .All these forces seem to be combining. . .to substitute. . .various forms of state control for exclusive copyright control. . .In this situation it is quite possible to envision the emergence of societies in which there is little individual or independent authorship." Ringer, *Copyright in the 1980s—The Sixth Donald C. Brace Memorial Lecture*, 23 BULL. CR. SOC. 299, Item 333 (1976).

[190] The compulsory license for public broadcast performances of certain copyrighted materials is an adjustment not properly technologically required, since blanket voluntary schemes by performing rights societies such as ASCAP and BMI have accommodated the need. The accommodation seems therefore to represent rather a public-policy reallocation of cost and access rather than a technologically compelled adjustment. *See* W. M. Blaisdell, *The Economic Aspects of the Compulsory License*, Study No. 6, prepared for the Copyright Office and Subcommittee on Patents, Trademarks, and Copyrights of the Senate Comm. on the Judiciary, 86th Cong., 2d Sess. (Comm. Print 1958).

[191] Even the compulsory license provision, §1(e) in the 1909 Copyright Act, for the recording and mechanical reproduction of music, was not the first such license in the American experience. Four of the state copyright statutes (those of Connecticut, Georgia, New York, and South Carolina) enacted at the suggestion of the Continental Congress between 1783 and

not, however, all of a kind: each addresses itself to a very different technological problem, and the justification of each varies in persuasive force. What they have in common is that the technological factors involved, by themselves embodying, if not creating, the copyright interests, have in turn been called upon to solve the problem of rewarding the author appropriately for them. We do not examine them separately and in detail here, but make two comments, one in general and one in particular.

In general, it is not clear that a government rate-making feature is a necessary part of a compulsory licensing scheme, which essentially deals with a problem of *access,* once Congress has determined that a particular use comes within the scope of the copyright scheme at all. But insofar as the government-determined royalty established by the rate-making mechanism (a combination of statutory fees and fees determined by a copyright royalty tribunal) approximates what the market mechanism might appropriately require—an intent for which there is support in the legislative history of the copyright act[192]—the underlying economic-incentive mechanism of copyright is not seriously disturbed. The principal modification of the author's "exclusive rights" is in control of access, an encroachment that in this context is arguably—and is seen by Congress to be[193]—more in the interests of orderliness and of reduction of transaction costs for all concerned than as a principled invasion of appropriate rights under the copyright scheme.

Cable television in particular posed difficult conceptual and practical problems, but Congress resolved them essentially within the framework of the copyright scheme by making two determinations. First, it drew a line with respect to secondary transmissions by making a profit-making distinction between a "passive" use and a "protected" area that, while not conceptually internally consistent, was nevertheless an *administrable* line.[194] Second, it managed to come up with an ad-

---

1786—*see* the New York statute at n. 39, *supra*—provided for a compulsory license when copies of a copyrighted book were not supplied in quantities and at a price judged reasonable by a court. *See* H. G. Henn, *The Compulsory License Provisions of the United States Copyright Law,* Study No. 5, prepared for the Copyright Office and the Subcommittee on Patents, Trademarks, and Copyrights of the Senate Comm. on the Judiciary, 86th Cong., 2d Sess. (Comm. Print 1956).

[192] *See, e.g.,* the discussion of the reasonableness of royalty rates in the Committee Reports on sections 115 and 116, and the Second Supplementary Report of the Register of Copyrights (1975), ch. IX.

[193] *Id.*

[194] *See* the discussion of § 111 in the Committee Reports.

ministrable scheme that balanced the general public-policy copyright interest and the governmental interest in the regulation of restricted communications channels. The scheme essentially confirmed the author's copyright interests, again principally by modifying control of access, this time by means of a three-level mechanism—(1) compulsory licensing by statute,[195] (2) rate determination by means of a royalty tribunal,[196] and (3) allocation of markets by a government agency, the FCC.[197] Complex as this solution at first sight appears, it constitutes on the whole a manageable accommodation of access wholly within the economics of the copyright scheme.

However one might see these compulsory licenses and passive exclusions as primarily technologically accommodative of the copyright tensions, the exemptions in Table II have to do more directly with substantive policy matters, often with the competing constitutional interests sharply posed.

3. Public Policy Exceptions.

The sheer number of exceptions identified in Table II as for public policy reasons (twenty-three), in comparison with the number identified in Table I as for technological reasons (twelve), would seem on the face of things to indicate that whatever might be the facts with respect to how much the United States copyright system has in the past consisted in fundamental and extensive "limitations and conditions" on the exclusive rights of the author, the new Copyright Act surely does. But since Table II includes exceptions as minimal as the playing of a phonograph record in connection with its sale, it is clear that a conclusion about the general shape of the new Act must await a weighing of the impact on the copyright scheme of particular provisions.

a. Peripheral Economic Adjustments. The six exceptions in Part A of Table II have here been categorized as "essentially not part of the usual commodity-market mechanism," and thus as having minimal impact on copyright-scheme economics. As such they do not bear heavily on the question of general copyright-scheme dynamics. There is nevertheless something important to our general purpose to be gleaned from this category: three of the six items have to do with the ambiguous way in which Congress has dealt with fair use, and contribute to the blurring of the fair use–exempted use line. That is, two

---

[195] § 111.
[196] §§ 801-809.
[197] § 111(c).

exceptions expressly set out as *exemptions* in the text of the statute—record playing by retail stores[198] and the use of a picture of a useful article in an advertisement or news report[199]—might more appropriately have been left to be excused under *fair use* doctrine, suppose it were more clearly defined; while another—reproduction of old motion-picture films—derived wholly from a statement in the Committee Reports' comment on the fair use section,[200] might rather have been expressly made in the text of the statute or, more consistently, be considered simply to be subsumed under the section, 108(c), dealing expressly with the preservation of library and archival materials from which another of the six is directly drawn.

If fair use as a doctrine is to be seen as encompassing uses not appropriately within the scope of an author's expected economic reward, both the playing of a record at the point of sale and the reproduction in an advertisement or a news report of a copyrighted design made expressly for a useful article expected to be sold in the normal ways of trade ought surely to be within the expected scope of any court's definition of fair use. If that is not so, the whole fair use notion is in trouble. But insofar as it *is* so, the clarity of the fair use notion is increased by the omission of such obvious "fair uses" from among the express exemptions in statute or report, and the line between exemptions and fair use rendered more obscure by the inclusion of them.

Similarly, absent a statutory section exempting a library from seeking permission of the copyright owner for making a copy of an out-of-print work, the notion of fair use would be under some strain, and again the exempted use–fair use line blurred, if the making of a duplicate of a copyrighted motion-picture film without permission of the copyright owner were naturally expected to be seen by any court as in a category of use falling clearly within the doctrine. But *with* an express section of the statutory text addressing the precise question directly elsewhere in the Act, the ambiguity about the fair use line occasioned by the committee commentary applies with even more force.

b. Integral Economic Adjustments. The sixteen exceptions in Part B of Table II have been categorized as "essentially accommodatable by the normal commodity-market mechanism" and thus as not only within the "appropriately expected economic reward" but fully capable of being accommodated, by normal trade practices or routine private agreements, within the copyright-scheme economic mechanisms.

---

[198] § 110(7)
[199] § 113(a) (2)
[200] Committee Reports, discussion of films in the commentary on § 107.

The range and the complexity of impact on copyright-scheme dynamics, however, is very great, and Congress has recognized this by treating two categories of these exemptions (those having to do with education and with libraries) that lie peculiarly at the center of the very nature of copyright—the precise use, furtherance of knowledge, for which the scheme is designed—in particular and intensive detail. It is accordingly appropriate that we treat separately the eight exceptions having to do with the use of books, records, and other copyrighted works in education and library contexts, and we do so below. First, however, we examine the remaining seven "integral economic adjustments" in Table II, which we here call simply "special cost reallocations."

(1) Special Cost Reallocations.

Of the seven exceptions in Section B of Table II not having to do with education or libraries, only one (the "fair use" making of a copy of a work for the blind) has to do with the making of a tangible copy of a work. The other six concern intangible but nonetheless real rights of performance and display. It is at once clear that however real the reallocation of costs among these seven uses, the accommodation of competing interests is either minimally disturbing of the copyright scheme, or, as in the case of the exclusion of performance rights in records from the scope of copyright protection, not incompatible with it. That is not to say that a question might not be raised about the clarity and rationale of one or another of these exemptions. For example, it is not on its face clear why a rationale for the exemption for music played at an agricultural or horticultural fair[201] would not also apply to an urban fair appropriately sponsored, or any number of other events of the same sort.

But except for the policy exclusion of performance rights in records for public as well as private use, the other exceptions are minor: the exemption for making a single copy for the blind is seen to be so minimal as to be subsumed under a fair use rubric; the two exemptions for religious performance of music and nondramatic literary works are narrowly circumscribed in time and place; and the "non-commercial" use of these materials for such arguably public benefits as performances for the blind and deaf are so near the margin of fair use as to raise no serious questions about cost reallocation.

It is clear that at this point in our analysis of the new Act's changes in the copyright scheme, the mere numbers of exceptions as

---

[201] § 110(6).

between technological accommodations and substantive public policy provisions are not determinative of the shape of the Act. So far, the more substantial changes have been technological. But however minor these separate public-policy exceptions, they do in the aggregate il-luminate the general and larger question with which we began our in-quiry into the rationale for exemptions, namely, that when the basic risk in and reliance on the internal dynamics of the copyright scheme have been taken, Congress takes control of access from the author for only two reasons: (1) because a competing constitutional interest re-quires that a particular use of the work should be *free of cost;* or (2) because a competing constitutional interest requires that access not be restricted *except for cost.*

These considerations are more sharply posed in the two matters that we consider next.

(2) The Educational Context.

The educational exemptions are expressly dealt with: (1) in the fair-use section (§107), (2) in the three sections dealing with displays and performances, including broadcasts (§§110, 112, and 118), (3) in the section on secondary transmissions (§111), and (4) in the section on remedies for infringement (§504), as well as being subsumed (5) in certain of the copying exemptions granted in the library reproduction section (§108) and (6) in the general display provisions of the section on the rights of the owner of a copy of the work (§109).

(i) Performances and Displays. With respect to performances and displays, it is clear from both the wording of the statute and the commentary in the Committee Reports[202] that the Congressional intent here is fundamentally to spell out the school-use exemptions that are subsumed in the express general "not for profit" exemptions in the act of 1909,[203] but extended to take account of new technologies, par-ticularly educational television. The careful limitations either on the kinds of works involved or on the conditions of instructional use re-flect a policy of keeping the exemptions more or less within the bounds of face-to-face classroom instruction. This is true even though three of the statutory accommodations of publicly broadcast educa-tional programs put some strain on those old boundaries. One of these (number 11) is the seven-year statutory license to make thirty copies (which may be used to perform these works) of certain educa-

---

[202] *See particularly* the committees' comments on § 110.
[203] Act of March 4, 1909 [Pub. L. No. 349], ch. 16, 4 Stat. 436, § 104 and § 1(e).

tional broadcasts,[204] and another (number 13) the statutory license to make ten copies of certain broadcasts for the handicapped.[205] The third (number 17) is the exemption that permits educational broadcasts capable of being received by the public at large.[206] In this last, the technical characteristics of the medium—uncoded transmission over the air—make of an otherwise controllable exemption something with the characteristics, so far as allocation of costs are concerned, of a "public good." This accommodation, however, as, indeed, all of the statutory rules with respect to educational displays and peformances, is essentially undisturbing of the dynamics of the copyright scheme.

(ii) Reproduction. In Part B of this chapter, dealing with the identification of exemptions in the new Act, we were interested, so far as education was concerned, merely in distinguishing notions of fair use—something within the anticipated dynamics of the copyright scheme—from exempted use, something expressly excepted from the copyright scheme. We found it necessary to define the educational exception for photocopying, to the degree set out in the statute and the legislative report, as being clearly an exemption rather than fair use. Here we are concerned with a different question: Is the scope of the exemption such as to disturb the dynamics of the underlying copyright scheme itself?

The incorporation in the legislative report of two "agreements on guidelines"[207] for school copying among several interest groups raises two preliminary questions: What is its effect on the doctrine of fair use generally? and What is the meaning of the agreement with respect to educational copying in particular? We deal with some aspects of the second of these first, noting particularly those provisions having to do with multiple copying, since those are most likely to pose the sharpest copyright-scheme questions.

---

[204] §§ 112(b) and 110(2).

[205] §§ 112(d) and 110(8).

[206] §§ 110(2) (C) (ii) and 118.

[207] House Committee Report, 67-72. One agreement has to do with books and journals, one with music. The agreement made among the ad hoc committee of educational institutions and organizations, the Author's League of America, and the Association of American Publishers dealing with books and periodicals is headed:

AGREEMENT ON GUIDELINES FOR CLASSROOM COPYING IN
NOT-FOR-PROFIT EDUCATIONAL INSTITUTIONS
WITH RESPECT TO BOOKS AND PERIODICALS

For the text of these guidelines *see* Appendix B, *infra*.

The "Agreement on Guidelines for Classroom Copying in Not-For-Profit Educational Institutions" raises the following questions, among others:

1. Since the number of permissible copies, defined by the formulation "one copy per pupil in a course . . . for classroom use or discussion," would allow, say, five-hundred copies to be made for a large college lecture class, or a thousand or more copies if the same lecture were given to two such groups, does the magnitude of the exemption pose copyright law questions as opposed to contractual questions?

2. With respect to the condition of spontaneity, which permits copying only "if it would be unreasonable to expect a timely reply to a request for permission," is it the intent that if there *is* time to consult, and the copyright owner says no, the teacher cannot then copy?

3. What is the time for a "timely reply"? Long enough to make a phone call? For an exchange of telegrams? An exchange of letters?

4. With respect to the two "spontaneity" factors "inspiration" and "decision," and to the time interval between those occurrences and the "moment of use," if "the [initial] inspiration" and "the [final] decision to use" are separated in time, does the time for obtaining permission begin with the former or the latter? Also, does the time for "maximum teaching effectiveness" include the time to get the work reproduced by the school system? If, say, the system takes forty-eight hours for the processing of a requisition and the photocopying, does the "unreasonable time" provision begin when the teacher submits the requisition? Or just before class, when the decision is finally made to use it? That is, with respect to "a timely reply" is the decision tied here to "use" or to "copy"?

5. With respect to "cumulative effect," what is a "course"? Is "course" restricted to instruction actually given by a single teacher, or does it include several "sections" of a teacher's class taught by teaching assistants at different times and in different classrooms? Is the answer different if all the instruction is given by a single teacher, but in several sections at different hours?

6. Though the "copying of the material" may be "for only one course in the school," may the student who has paid for the material *use* it in another course?

7. What is the "actual cost of photocopying" for which the student may be charged? Besides the cost of paper, what fixed costs are part of the "actual costs"? The yearly lease-cost of the copying mechanism? The cost of personnel operating the machine? Maintenance costs? Overhead allocations?

8. With respect to the condition that copying not be a "substitute for the purchase of . . . publishers' reprints," if a publisher has and

sells reprints, and the teacher does not know this, is he excused from this section? If he does know that reprints are available, would he nevertheless be excused by virtue of the "spontaneity" and "timely reply" provision?

9. In view of the requirement that the copying be done only "at the instance and inspiration of the individual teacher," is the condition that the copying not be "directed by a higher authority" merely redundant, putting in negative form what the spontaneity section puts in positive form; or do concepts like "directed" and "higher authority" modify the conditions? Is a teachers' curriculum committee "higher authority"? If something is discussed by a teacher in such a committee and given approval, is it thereupon "directed"? If the suggestion is made by another teacher, is its use barred?[208]

10. Does the condition barring copying of the same item "from term to term" bar a teacher from skipping a term and then using the same piece?

11. What is the relationship between the limitation in the heading of the agreement to "not-for-profit educational institutions" and the statement in the House Committee Report that the revised wording of section 107 "is not intended to be interpreted as any sort of not-for-profit limitation on educational uses of copyrighted works"?

---

[208] The spokesman for the House Judiciary Committee, in presenting the copyright bill to the House, appears to deal with these questions in a gloss on the meaning of "individual teacher" and "higher authority" in the Guidelines:

". . .in planning his or her teaching on a day-to-day basis. . .an individual teacher will commonly consult with instructional specialists on the staff of the school, such as reading specialists, curriculum specialists, audiovisual directors, guidance counselors, and the like. As long as the copying meets all of the other criteria laid out in the Guidelines, including the requirements for spontaneity and the prohibition against the copying being directed by higher authority, the committee regards the concept of 'teacher' as broad enough to include instructional specialists working in consultation with actual instructors." 94 Cong. Rec. H-10875 (Sept. 22, 1976).

But these remarks raise two troublesome questions. First, if this gloss in fact now governs the agreement, it would appear to undermine the basic rationale for the educational exemption—namely, the requirements of spontaneity in the classroom. It is not clear what, if anything, is in practicable terms left of a "higher authority" test distinguishable by a teacher or administrable by a school. Second, since the agreement is an arrangement among three private groups, it is not clear what weight should be accorded the House Committee's view of what the contracting parties meant by one of its central terms, or what the effect would be of one or another of the parties' disagreeing with the Committee's interpretation.

The variety and extent of these questions, the tentative language of the purpose in the preamble of the agreement, and the ambiguous authority of the membership associations signing the agreement—all raise the more serious questions of the scope of the educational exemption and of the fair use section generally. Who in the eyes of a court is governed by the agreement? Does it matter whether or not a particular publisher, school district, or college belongs to one of the signatory organizations? Whatever the meaning and force of the agreement itself, what is the effect on the copyright law of the inclusion of the agreement in the legislative report? The House Committee Report says of it, "Teachers will know that copying within the guidelines is fair use." Whatever the intent of Congress with respect to how a court is to be guided by the terms of the agreement as set out in the Committee Report, are there notice problems here— particularly if changes are subsequently made by the parties to the agreement?

However these questions might be answered, a court in a particular case would be governed by section 504 (c) (2), which deals with liability standards and with remedies for infringement. Since the legislative history of section 504 (c) shows that Congress intended to give teachers broad latitude in interpreting section 107 of the statute,[209] courts would be justified in taking the most liberal view of the educational exemption. And since the fair use section cites particular uses only as examples[210] of how the doctrine is to be applied, courts would be equally justified in applying the educational standards to analogous uses.

There is some reason to wonder whether the educational exemption, thus dealt with by Congress under expansive and ambiguous notions of fair use, is administrable by courts. More serious, however, is the strain put on the dynamics of the copyright scheme. The fierce debate on the educational exemption issue may have led Congress to what might be the worst possible outcome: neither squarely facing the exemption question so as to establish with precision in the statute itself a set of administrable rules reflecting an appropriate level of constitu-

---

[209] The House Committee Report on § 107 says that in an effort to meet the "need for greater certainty and protection for teachers. . .the Committee has. . .amended section 504(c) to provide innocent teachers and other non-profit users. . .with broad insulation against unwarranted liability for infringement."

[210] "The examples enumerated at page 24 of the Register's 1961 Report, while by no means exhaustive, give some idea of the sort of activities the courts might regard as fair use under the circumstances." House and Senate Committee Reports on § 107.

tional balancing, nor relying on those fundamental workings of the copyright scheme that rest either on the mechanics of the marketplace, including voluntary contractual relationships, or on the discernible "appropriateness" of the author's "expectations of economic reward." How much, if at all, this is fundamentally disturbing of the copyright scheme, we shall consider after an examination of a copying exception that Congress *did* try to encompass in an express exemption in the statute, namely, library photocopying.

(3) The Library Context.

The library photocopying question, which involves principally the exemptions 2, 3, 4, and 5[211] in our catalogue, has aroused no less debate than educational photocopying. The fundamental questions at issue, however, are of a very different sort. Not only has educational photocopying to do with very small bits of copyrighted works, used spontaneously in a temporary mode, but it involves the actual use of the work by the photocopier himself (a teacher in his institutional role) rather than by an intermediate service agency (the library). That is, because the copier's use in the teaching context is direct—both in the sense of the mind's "possession" of the intellectual work of the author by its ultimate perceiver, whether teacher or student, and in the sense of an institution's employment of a particular work for particular institutional purposes—the copyright-scheme relationship between creator and user has never been obscured, however exacerbated the argument about the appropriateness of the author's economic reward. The central question, that of *promptness of access* in a particular situation—classroom teaching in "face-to-face" activity—has been clear: the "spontaneity" criterion *governs* the exemption.

Libraries, unlike teachers or students (it is not the *schools* to whom the educational photocopying privilege is given), do not read books or listen to records, or otherwise "use" such works themselves for their intrinsic purposes. They are—or, at least, traditionally have been—simply agents of users. And such users have all of the varied characteristics of the ordinary purchaser of a work in the marketplace: a wide range of choice, of priority of resources, of urgency. That is, a library represents a collective endeavor, in which a group of disparate users, whether individually or as part of a using institution, shares the costs of selecting, purchasing, and housing books, periodicals, records, and other works. In that sense, the relationship in library photocopy-

---

[211] *See* text at notes 128-137, *supra.*

ıng is like that between the individual user and the author—the usual copyright-scheme economic relationship.

Typically, too, the library takes what appears to be a natural place in the market for commodities of the copyright scheme. Though the effect of the library-as-agent logically implies the substitution of a group purchase for a number of individual purchases, it is normally not seen by any of the actors in the copyright scheme as anything but an undisturbing and fully accommodated factor in copyright-scheme economics. Indeed, beyond their multi-faceted role as facilitators of access to knowledge, literature, and the arts, and as catalyst in the intricate and arcane process by which individual users buy copies for themselves, libraries are typically seen as constituting an essential part of the market itself.

But the characteristic of a library-as-agent poses difficult conceptual problems with respect to an appropriate copyright-scheme approach to photocopying. Whether Congress has in the end dealt with it in a coherent way we have now to consider.

We begin by setting out some of the primary provisions of section 108.[212]

(i) The library as agent.

With respect to an inquiry by a patron, what the library exemption does is provide for the following:

(1) A "library or archives" that is open either to the public generally or to anyone "doing research in a specialized field" is given a qualified right, to the exclusion of any other reprinting agency,[213] to reprint and to distribute without the permission of the copyright owner and without the payment of royalties, provided there is no "purpose of direct or indirect commercial advantage":

    (a) a complete copyrighted work, provided it cannot be obtained at a "fair price,"[214] and

---

[212] The text of Section 108 is set out in Appendix A. *See also* the House Committee Report, 74–79, and the Conference Report, 70–74. The CONTU guidelines on library photocopying appear in Appendix C, *infra.*

[213] If one of the policies of § 108 is to protect the copyright owner against dilution of the exclusive-right market by libraries, § 602(a) (3) seems curiously to cut the other way by permitting the importation of five copies or photocopies of a copyrighted work legally made abroad so long as they are ordered by a library for its "library lending or archival purposes" and so long as they *otherwise* do not violate section 108(g) (2). *See also* the text at n. 219, *infra.*

[214] It is not clear what "fair price" standard would apply if: (a) the publisher itself offered to provide a photocopy at a cost reasonably equivalent to

(b) an article in a periodical, a contribution to a collective work (such as a paper or chapter), or a "small part" of any work, irrespective of whether reprints are available from the author or publisher,[215] provided one or another of some complicated rules about multiple copying is not violated.[216]

(2) The library need not do the reprinting itself, but may act as an ordering service for the patron so long as the copy is provided by another similar agency.[217]

(3) The library need not have a copy of the work, nor need ever have had a copy of the work.

(4) But if it does have a copy of the work, the patron need not borrow it in order to use it. The decision about whether to borrow it or to have the library make him a copy or to make it himself—and the responsibility for the legality of the copying—is his.[218]

(5) There is no provision in section 108 for the library's making for *itself* a copy of a copyrighted work it has never owned.[219] The library either: (a) must give the copy to the person ordering it, either free of charge or at a cost that satisfies the test of being "without direct or indirect commercial advantage" either to the reprinter, or, if it is not the reprinter, to the distributing agent,[220] or (b) may keep it, provided it is only replacing a damaged or missing copy in its own collections.[221]

(6) In processing an individual order for a copy of all or part of a copyrighted work, the library must be unaware ("have no notice") that

---

what it would cost the library to make it itself; or (b) the publisher offered either to provide such a copy or to grant to the library permission to make its own with, say, an appropriate royalty based on the library's photocopying costs, appropriately defined. *See also* the text at n. 238, *infra*.

[215] The statute is silent on the question so far as library copying under § 108 is concerned, as is the commentary in the Committee Reports. *But see* the text at n. 207, *supra*, with respect to publishers' reprints and copying by teachers, where the question is perhaps answered differently under § 107 guidelines.

[216] § 108(g). *See* text at n. 233, *infra*.

[217] *But see* n. 213, *supra*, on § 602.

[218] § 108(f).

[219] §§ 108(d) (1) and 108(e) (1) require that "the copy [become] the property of the user." There might be one exception to this in another context: subsection 108(f) (3) permits the making and presumably retaining of "a limited number of copies. . .of an individual news program," an exemption akin to those exempted as "ephemeral recordings" under section 112 of the statute.

[220] § 108(a) (1). *See also* the text at n. 251, *infra*.

[221] § 108(c).

the customer intends to use it for "any purpose other than private study, scholarship, or research."

The statute appears, particularly in the last proviso enumerated above, to reflect with some precision both the general exclusive-right control of the author in making copies explicit in the copyright scheme and the implied fair use exception for personal use in what might be called the "closet" mode of private use. At least section 108 starts out that way. There are, until subsection 108(g) (2) is reached, four separate statements, two express and two implied, of the governance, to the exclusion of any other rationale, for the *exemption* (as opposed to the way the exemption is implemented), of individual private use: Subsections (d) (1) and (e) (1) expressly require that the use be for "private study, scholarship, and research;" subsection (f) (2) invokes the same phrase by its reference to the fair use standards of section 107; and subsection (g) (1) implies the same standard in phrasing the notion in exclusionary terms.

At first sight, the copyright scheme appears undisturbed, with both the library and the individual user playing traditional roles. But on close examination a subtle but important change in the traditional relationship seems to have occurred.

The traditional model of the library as set out above characterizes it as an agency that purchases works, in accordance with the interests of its patrons, its acquisitions policy reflecting those patrons' priorities with respect to subject matter and breadth of coverage, and its overall dimensions reflecting the allocation of group resources between the library and other group needs. Typically, once a work is purchased it may be consulted or borrowed by a member of the group without further direct cost. The work becomes a sort of "public good" within the context of the library, with housing and lending transaction costs not allocated by particular book or particular patron.

The provisions of section 108 outlined above seem explicitly to reflect that model, in that in the making and distributing of copies of a work the emphasis is not on the library, but on the user. As with any intellectual work, the "user" is an individual, and as in all copyright-scheme economics it is the *user's* need to see or hear the particular copyrighted work that is the stimulus for the consequent economic transaction involving incentives and economic reward. The statute, in requiring that the user, rather than the library, own the copy that is made of a work, appears to confirm the traditional reliance on the balancing *inherent* in the copyright scheme. But though the library is still an agent in the scheme, *its role has somehow been changed by section 108 from that of agent of the user to agent of the author.* That is, the library, rather than buying a book or periodical on behalf of a group of users, simply provides the mechanism by which an individual user buys the work for *himself*—either by the library's reproducing the work on the premises with its own equipment or by the library's providing the order-taking office to have it reproduced elsewhere.[222] The library, in this configuration, is simultaneously the selling agent, the manufacturing agent, and the collection agency.[223]

How this has come about requires an examination of the traffic involved in interlibrary transactions, and of the provisions of subsection 108(g) (2) that deal with multiple copying, particularly the "provided" clause and the extra-statutory "guidelines" generated by it.

(ii) Interlibrary transactions.

If a patron wants a work not available in his library, there are three possible solutions not involving the purchase of a copy outside the usual copyright scheme: (1) he purchases it himself, (2) the library purchases it for the group, or (3) the library borrows it from another library. The interlibrary loan route involves the library in certain transaction costs, but it avoids the cost of buying the work for its own collections on behalf of its patrons. It does so either because the work is unavailable for purchase or because the work is not within the group's priorities for acquisitions. There is, however, usually no charge to the user. Also, the cumbersomeness, the inefficiency with respect to transaction costs, and the time-consuming nature of the transaction are enough discouraging both to the user as potential purchaser of the work and to the library as potential agent-purchaser of the work to keep the use of the interlibrary-loan device ordinarily

---

[222] There is of course a third way for the patron—he can borrow the book from the library and take it either to his own copying machine or to that of any copying service—but that has nothing to do with the library's role as the copier under the statute, our only concern here.

[223] There is also with respect to a journal the possibility that the library may be seen as a sort of publisher as well. Something more than single-article reproduction may be happening when a journal is thus fragmented into its constituent parts. If a journal is a mechanism by which a collection of different articles is marketed as a unit to an audience that accepts a leveling of the degree of interest in each reader in a particular article in order to share certain costs of editing, manufacturing, and distribution, when the audience is redefined as constituting a single-article market, there may no longer be the same economic rationale for the publisher's particular amalgam: each redefined market in effect groups articles of various sorts to suit its narrower clientele. In short, if cost spreading is going on, it would appear to be by the library as "editor" and "publisher" by virtue of its service function.

within bounds that leave copyright-scheme economics essentially undisturbed. That is, it is normally quicker and perhaps cheaper to buy a work that is likely to be used more than once.[224]

It is not clear that any of these "inconvenience" factors affecting *access* to a remote source of a work is significantly altered by the availability of photocopying, but there is no doubt about two substantial "tangibility" differences from the former mode: a new copy of the work is permanently made, and there is a statutorily required[225] shift of some of the costs from the library-as-agent to the patron as consumer. To see what effect these changes might be expected to have on the copyright-scheme dynamics it is necessary to distinguish in the new photocopying mode two different relationships between patron and library.

The first is the relationship based on the analogy to the private-use handcopying, where the patron instead photocopies a reasonably small part of a work that his library has in its own collections. The difference from handcopying is only that instead of the patron's incurring no further cost, he pays for the product of the copying machine as a convenience to himself. The library has no direct role in the matter. The patron always has the option of just reading the work in the library, or borrowing it, or hand-copying it free. His new cost, that is, insofar as the reproduction is in the hand-copying mode, is completely *outside* copyright-scheme economics.

The second library-patron relationship is analogous to the ordinary marketplace transaction that takes place whenever there is a routine sale of a copyrighted work. The patron buys a copy, though here through the agency of his library, as he would in other circumstances through any vendor. His new cost here has all the appearance of something normally *within* copyright-scheme economics.

The consequence of this second relationship is at least that the

---

[224] An estimate is made in ANNUAL REPORT OF THE STANFORD UNIVERSITY LIBRARIES, 1975-76 (Stanford University, 1976): "A realistic look at circulation costs indicates that it may cost 10¢ to circulate a reserve book, 20¢ from the general stacks collection, $1.00 from the locked stack, and $2.00 from a campus auxiliary stack collection, but it costs from $4.00 to $9.00 by interlibrary loan."

And if the interlibrary transaction were to involve photocopying there are presumably additional costs. In 1976, for example, the Stanford Library charges to another library for copying parts of a thesis were 15 cents a page, plus costs of searching, postage, and handling. For a 50-page extract, for example, an additional cost of $7.50-$10.00 would appear to be involved.

[225] *See* n. 219, *supra.*

"interlibrary loan" notion has been subtly altered. It is clear that many of the characteristics of an interlibrary loan that in pre-photocopying days made it essentially undisturbing of copyright economics have been replaced by something that resembles the ordinary marketplace transaction mode. The transaction is neither a loan between libraries of something to be returned nor a cost-free loan by the library to its patron. The characteristics that make the remote duplicating system part of a vending mechanism requires us in due course to consider how to look upon its effect as a substitute for the purchase of a work either by a library or by the individual through normal channels, and what, if anything, to make of the fact that the copy is provided to the user by means of a purchase of the work *outside* the copyright scheme.[226]

(iii) The library's policing role

In the sequential exposition of the library's photocopying role under section 108, subsection (g) (2), particularly the proviso clause and its progeny, the "CONTU guidelines," represents a sharp alteration of focus. Until that subsection is reached, the library's role as described in section 108 is essentially that of a neutral agency, with service and policing functions responsive wholly to the needs of the individual user. The library's responsibilities with respect to the patron's use of the photocopying can heretofore be given as seven:

(1) to give patrons appropriate warnings about infringement of copyright;[227]

(2) to include a copyright notice on any work the library reproduces for patrons or for itself;[228]

(3) when a work is out of print, to make the determination that it "cannot be obtained at a fair price";[229]

(4) to transfer ownership of the photocopy to the patron;[230]

(5) to refuse to supply a photocopy for a patron if it has "notice

---

[226] This applies as well to a journal article or chapter in a symposium volume as to a complete book or journal, for the photocopying mechanism that poses the problem also makes parts of works even more separately vendable: the vendor of the photocopy does not have to decide in advance of marketing which parts to print in quantity, or, in some cases, how many.

[227] § 108(d) (2) and (e) (2).

[228] § 108(a) (3).

[229] § 108(c) and (e).

[230] § 108(d) (1) and (e) (1).

that the copy could be used for any purpose other than private study, scholarship, or research";[231]

(6) to refuse to supply to a patron a copy of more than "one article or other contribution to a copyrighted collection or periodical issue . . . or . . . a small part of any other copyrighted work",[232] and

(7) to refuse to supply a photocopy if the library "is aware or has substantial reason to believe that it is engaging in the related or concerted reproduction or distribution of multiple copies or phonorecords of the same material, whether made on one occasion or over a period of time, and whether intended for aggregate use by one or more individuals or for separate use by the individual members of a group."[233]

So far the rules appear to be coherent, in that each of them has to do with the library's service role and policing responsibilities in responding to the orders placed by individual patrons. Section 108(g) (2), however, has a different focus altogether: instead of the individual user's purpose and intent, the statute here is concerned with the *library's* purpose and intent.

Since there is no room in the internal balancing of the copyright scheme for a straightforward and substantial market exception that is clearly within the "appropriately expected economic reward" of the author, the accommodation can be rationalized only by redefining plain words. That is, the statute by talking about "such aggregate quantities as to substitute for a subscription to or purchase of" a work begs two questions—first, whether the traditional interlibrary loan is not in fact a substitute for a purchase by the borrowing library, and second, whether even one photocopy is not, in fact, "a substitute for a . . . purchase of a work." Of course it is. Of course both transactions are. By falling into this conceptual trap, Congress then required a rule by which the ordering of an arbitrary number of photocopies of articles in periodicals or contributed papers in a collective work was defined as not constituting a substitute for purchase either individually or in the aggregate. The CONTU guidelines define the number as 25 for a periodical in the first 5 years of its publication, at a rate of no more than 5 copies a year, and perhaps no limit after 5 years; and for contributions to "collective works" (i.e., books) as 5 copies a year without limit of time.

However uneasy Congress might have been with this solution— and the requirement in section 108(i) that the thing be looked at again

---

[231] § 108(d) (1) and (e) (1).
[232] § 108(d).
[233] § 108(g) (1).

in five years reflects that unease—it is the failure to focus clearly on the first conceptual problem, the benign but potentially troublesome nature of the traditional interlibrary loan, that might, in the end, be the source of the flaw, if there is one, in the statutory solution. The dubious and logically contradictory nature of the proviso clause reflects an unhappy conjunction of the conceptual confusions that Congress had separately more often than not managed to avoid during the long years of gestation of the new Copyright Act—the nature of the basic copyright-scheme risk, the complex agency nature of the library, the peculiar identity of photocopying with the initial mode of producing the copyrighted commodity, the internal dynamics of copyright-scheme economics, the administrability of the exemption. It is a coherent view of these factors that must, in the end, determine whether the Congressional solution to the photocopying problem is sound and workable. We shall look at that in Chapter V.

But whatever the success of the arbitrary "guidelines" definitions imposed on the inherently self-contradictory standard of section 108 (g)(2),[234] no such procrustean solutions settle other problems of section 108 that are just as troublesome. We examine some of these in Chapter IV.

---

[234] The discussion here does not end the questions that the CONTU guidelines raise. While the Conference Report's disclaimer ("the guidelines are not intended as, and cannot be considered, explicit rules") gives courts some leeway, the legislative force of the report, and the "shall" terms in which the guidelines are phrased, will almost surely govern what a given litigant might do. In that context, the questions raised by a close examination of the CONTU text are not academic.

To begin with, since the statute bars photocopying by one library for another if the arrangement has "as [its] purpose or effect" the substitution of the photocopy for "a subscription to or purchase of [a] work," what is a court to do, if, despite the CONTU definition of what constitutes a substitution for a purchase, two or more libraries *agree* that such is the "purpose and effect" of even a single purchase? For example, one purpose given for the establishing of a cooperative interlibrary program between Stanford University and the University of California at Berkeley was to coordinate acquisitions of books and journals so that instead of each institution's buying a work, only one would. The Times, Palo Alto, Calif., Nov. 9, 1976, p. 25. If, now, photocopies are involved in implementing this interlibrary arrangement, is the wording of the statute or the definition of the CONTU guidelines to govern the question of whether there has been a "substitute for a subscription to or purchase of [a] work"?

And within the guidelines themselves there are problems. For example, while the central thrust of the guidelines, dealt with in paragraph 1(a), has to do with copyrighted material *not* owned by the library (and therefore clearly raising the substitute-for-a-purchase question), another part, paragraph 2(b), has to do with works already owned by the requesting library but that the library wants copied by another library from *its* collections. The standard is whether an owned work is "not reasonably available for use" at the first library. The "availability" may be something that may or may not be temporary, but it does raise a question about whether the photocopy is a "substitute for. . .a purchase" of the work by a library in a context different from paragraph 1(a). For paragraph 2 of the guidelines, by invoking the copying provisions of the "other provisions of section 108," which have nothing to do with the "substitute" question, appears to take it outside the five-copy limitations of paragraph 1. Does section 2(b) mean that no matter what the needs of its patrons, a library need never purchase more than a single copy or subscription of a work? In any case, it would clearly under paragraph 2 of the guidelines be possible for more copies of journal articles and of contributions to collective works to be photocopied for a library than are defined by the guidelines under paragraph 1 as constituting a "substitute" for a purchase or subscription.

IV.

The Photocopying Adjustments in Copyright Scheme Dynamics

The examination in Part C of Chapter III of some basic conceptual problems—particularly the societal and copyright-scheme role of the library and the meaning of the proviso clause of section 108(g)(2)—makes clear without more that the photocopying exemptions embodied in the new Copyright Act have ramifications beyond a consideration of the practicalities of library-service copying. But before we go on, in Chapter V, to consider the overall question of the Copyright Act's approach to exemptions in general, it would be useful to consider some of the problems of interpretation and implementation of section 108 that affect the administrability of the scheme, for that will in turn have some bearing on the larger issues. We examine six questions here.

A. "Obtaining" a Copy

Subsections 108(c) and 108(e) permit reproduction of an "entire work . . . or . . . a substantial part of it"[235] if an unused[236] copy "cannot

---

[235] The phrase "or to a substantial part of it" appears only in section 108(e), which has to do with copying an "unavailable" work for a patron ("user"); the phrase does not appear in section 108(c), which permits a library to copy a work for itself as a "replacement" for a copy "damaged, deteriorating, lost, or stolen," but clearly the library's right to make a copy of the complete work for itself includes the right to make a copy of a "substantial part of it."

[236] The word "unused" appears only in subsection 108(c), but again it is clear that the adjective is implied in subsection 108(e) as well—that is, it is a new

be obtained at a fair price."[237] The statute requires a "reasonable investigation"[238] about the availability of an unused copy, but the wording of the statute raises a complex question about the meaning of "availability" in the copyright-scheme design that is not expressly addressed by the statute. Suppose the copyright owner does not have a copy of the book in stock and has either not licensed any reproducing service to make copies or has issued such licenses on a nonexclusive basis, but is willing to grant permission to the inquirer to make a copy for a fee. Whatever might have been the legislative intent with respect to the meaning of "cannot be obtained," the availability of a work by means of a one-time license to reproduce a work would seem to be within the implied definition of obtainability. For if there were a question from the inquirer about the amount of the fee, all the copyright owner need do is to offer to make a photocopy for him. The inquirer could choose to accept either the copy offered or the license to copy, but so long as the copyright owner does not refuse both—something the copyright act does not allow him, though the British act, for

---

copy, not a secondhand copy, that need not be "available" if the user is to take advantage of the 108(e) exemption. All doubt of the Congressional intent here is removed by two indications in the commentary on subsection (e) in the Committee Reports: first, the commentary uses the phrase "unused copy" in paraphrasing the statute ("a reasonable investigation to determine that an unused copy cannot be obtained will vary"), and second, the paragraph on subsection (e) is headed "out-of-print works," though the following text does not itself use either the term or a euphemism for it.

[237] The meaning of "fair price" is discussed *infra*; *see* text at n. 240.

[238] Subsection 108(c) uses the term "reasonable effort," subsection 108(e) "reasonable investigation," but the commentary in the Committee Reports uses only "reasonable investigation" in the discussion of both subsections. The description of the effort to be made is identical in the Committee Reports' discussions of both subsections: "It will always require recourse to commonly-known trade sources in the United States, and in the normal situation also to the publisher or other copyright owner. . .or an authorized reproducing service."

A complicating factor for a court is language in the Senate Committee Report on section 107, on fair use, which in its discussion of the "availability of the work" says, "If the work is 'out of print' and unavailable for purchase through normal channels, the user may have more justification for reproducing it than in the ordinary case, but the existence of organizations licensed to provide photocopies of out-of-print works at reasonable cost is *a factor to be considered*." [Emphasis added.] This would appear to swallow up the section 108 options with respect to photocopying out-of-print works, particularly since section 108(f) (3) says that "nothing in this section. . .in any way affects the right of fair use as provided by section 107." It is unclear whether the term "reasonable cost" is descriptive or prescriptive.

example, does[239]—the "availability" of the work is arguably not in doubt. What then remains at issue is the meaning of "fair price," a question that as well raises other questions.

B. "Fair Price"

There are three situations in which the question of whether a price is fair may arise: (a) the normal situation, in which the seller means to sell as many copies as he can at normal market prices; (b) the "rare book" situation, in which a dealer has a stock of unused copies of a work that now commands premium prices;[240] and (c) the denial-of-access situation, in which the copyright owner's intent is to prevent the distribution of a work by putting a prohibitive price on it. The unused rare-book situation need be no more than mentioned here, for whatever the problems raised by it, it would be so infrequent as scarcely to be disturbing of copyright-scheme economics. The other two, however, could pose difficult questions for courts.

It is unclear why Congress has imposed the "fair price" test at all, rather than simply an "unavailability" test, as the standard for triggering the reproduction exemption. But since the test is in the statute, it must be applied by courts, and accordingly we must consider what factors might be applicable in determining the fairness of a price. To begin with, there is no sign, in the statute, in the legislative report, or in the hearings, that Congress in fashioning the complete-work copying exemption considered the various situations, other than where a work is traditionally "out of print," in which the fairness of a price could be at issue. The text of the statute does not in fact use the phrase "out of print" at all; the term appears only as the heading to the Committee Reports' discussion of subsection 108(e)[241] and in the discussion of section 107 on fair use.[242] But neither the statute nor the reports require the application of the out-of-print test to any work, in any circumstances. Accordingly, two difficult and troublesome

---

[239] Section 7(5) (b) of the British Copyright Act makes it a condition of the library's photocopying exemption that "the library. . .does not know the name and address of any person entitled to authorize the making of the copy." Copyright Act, 1956, 4 & 6 Eliz. 2, ch. 74. In short, the British statute in the ordinary situation *requires* the copyright owner's permission to photocopy.

[240] An "unused book" could be available as a "rare book" as well if a publisher had disposed of all remaining (and substantial) stock of an edition to a dealer, who still held them after their value had markedly increased.

[241] *See* n. 236, *supra.*

[242] *See* n. 238, *supra.*

questions are finessed. The first involves a work that is in print through regular channels of the book trade, but that the user or library might consider unfairly priced. It is to be guessed that a library might consider an "in print" classification of a work as presumptively determining that the price were fair, whether or not the user-purchaser agreed. And a court would more likely than not accept that presumption. The second finessed question has to do with the definition of "in print." The legislative report seems at first glance to have addressed it, however obliquely, by its invoking the "authorised reproducing service" in connection with the sort of "reasonable investigation" of the availability of the work required. But the "fair price" test, which must go beyond mere availability if it means anything at all, casts the presumptive force of the mere existence of an "authorized reproducing service," or any other licensed reprinting agency, in doubt, and triggers the whole range of possibilities having to do with what an inquirer considers a fair price.

In the "out of print but available on demand" situation, if there is a difference of opinion between the user and the library about what price is fair, whose judgment governs? Must a library photocopy at a user's request if the library "determines" that the price from the copyright owner is fair but the patron does not agree? May the user "shop" for a library whose view agrees with his, or is he bound by the determination of the first library he asks? But supposing the "normal" and the "in print" situations are ultimately resolved by the courts, how is a court to distinguish between "normal" situations and the "denial-of-access" situation? That is, whatever standards of availability and of costs the court may in due course set, what weight should a court give to a copyright owner's interest?[243]

Suppose, for example, that a copyright owner wished to bring an out-of-print work back into print at a price—say, $15.00—that would insure its widest distribution in a normal market; that the book could be expected to sell only 50 copies a year; that a printing of 500 copies were required to obtain a cost allowing a price compatible with an annual sale of 50 copies; but that the cost of capital and other con-

---

[243] The Committee Reports' commentary on fair use with respect to the right to make copies of out-of-print works, see n. 238, supra, expressly invokes the concept of the availability of a work through "normal channels" as a test triggering the possibility of permitting the copying of the work. Even the existence of a licensed "on demand" reprinter is only "a factor to be considered," not a definitive bar to making a copy under the fair use doctrine.

straints require that the reprinting be sold out within five years. If in the interim he sets a price of $50.00 on an on-demand reprint so as to hold the market for a few years until the work can be reprinted and sold in sufficient quantity to permit a $15.00 price, is the price "fair"? Suppose a library can photocopy the work at a cost of $30.00 for photocopying and paper alone, without fee; may it do so? Suppose the copyright owner has no intention of reprinting the work but sets a $50.00 price on an on-demand copy supplied by itself anyway? Or offers a one-time single-copy reprint license for $10.00? For $20.00?

It isn't clear how a court should decide such questions, in the long-term public interest, but these questions do indicate the hazards of statutory intrusion into the internal dynamics of the copyright scheme. For what section 108 does is to put the court back in the "access" and "price control" business (albeit with respect only to copies produced one at a time) that obtained for a while in England under the Statute of Anne and in some of the states under the copyright acts passed during the period of the Articles of Confederation.[244] That is, since the patron could presumably be served by means of an ordinary interlibrary loan of the traditional sort, the question of whether the exemption applied would necessarily rest not on access but on the fairness of the price. At the very least, it is clear that the misleading "interlibrary loan" characterization of the role of the library as purveyor of reprints to patrons obscures an intrusion of substantial dimensions into the copyright scheme. A court in judging the fairness of price might also, then, have to take into account the reasonableness of the patron or library in reproducing rather than in fact borrowing the work from another library.[245]

---

[244] *See* text at n. 38, *supra*. The British statute leaves the Statute of Anne where it found it and makes no mention of price, reasonable or otherwise. When a book is not available in England, the copyright owner must be asked, as in the United States statute, but if he denies permission to make a copy of the copyrighted work, that is the end of it: the library may not copy at all. Copyright Act, 1956, 4 & 5 Eliz., ch. 74, § 7(3).

[245] Section 504(c), which deals with remedies for infringement, provides for complete remission of statutory damages if a library or library employee "believed and had reasonable grounds for believing that his or her use of the copyrighted work was a fair use under section 107." But such complete remission of statutory damages is available to no one if the infringement, however innocent, has to do with section 108. That is, a failure to abide by section 108 requirements (for example, by failing to make a "reasonable investigation" of the availability of a work by writing the publisher) or a misjudgment of a fact (such as wrongly deciding, in the court's judgment, that the price of a book was too high), while permitting substantial mitigation of statutory damages, requires minimum damages of $100.00.

C. "Later Use" Liability

A long-standing provision of copyright law, embodied in the new act in section 109, is that, in the words of the Committee Reports, "where the copyright owner has transferred ownership of a particular copy . . . of his work, the person to whom the copy . . . is transferred is entitled to dispose of it by sale, rental, or any other means . . . [T]he outright sale of an authorized copy of a book frees it from any copyright control over its resale price or other conditions of its future disposition."[246] How far does this principle apply to photocopies? Since a copy of a work made under the photocopying provision of the act would by definition be "lawfully made," it is not clear what standards a court should use when a library patron sells the copy initially (and lawfully) made for his "private study, scholarship, or research."[247] Does section 108 bar him from subsequently selling it? From giving it away? If not, how long must he keep it? Is the answer different as between selling and donating the copy?

The answer is not made clearer by a confusing reference to the problem with respect to copies of articles in periodicals and to contributions to a collective work permitted under subsection 108(d). Subsection 108(f)(2) states that no one is excused from liability for copyright infringement: (1) if the making of a copy under subsection (d) "exceeds fair use as provided by section 107" or (2) if the "later use of such copy" exceeds such fair use. We skip over the puzzling ambiguities raised in the first part about whether this exemption or fair use permits the more extensive copying, and focus on the clearer question—however unclear the answer—having to do with the "later use" of the copy.

Does this language merely make explicit, with respect to articles and parts of books permitted to be reproduced under section 108(d), what we have inferred had to follow from section 109 with respect to the disposal of any copy of any work? Or is it meant to be restricted to works reproduced under subsection (d) (dealing essentially with articles) but not under subsection (e) (dealing essentially with books)? And, if so, what are the answers to the questions above about post-

---

[246] House and Senate Committee Reports, commentary on § 109. This principle would appear to nullify the effectiveness of wording sometimes printed on the copyright page of a paperback book that attempts to bar the rebinding and reselling of the book as a hardbound work.

[247] The library, having made the copy lawfully, has no liability for a patron's subsequent use of it. If there were any doubt about this, the commentary in the Committee Reports on subsection 108(f) makes the intent of Congress explicit.

private-use disposition of such copies? Are standards for "later use" by the copier different for articles and books?

Again, once one gets beyond the sort of private-use copying that fair-use notions would permit, difficult copyright-scheme disturbances occur that are not resolved by the talismanic invoking of the term "fair use" in an exempted-use context. The same sort of obliquely recognized contradictions reveal themselves here as appear in the ambiguities surrounding the use of the term "inter-library loan" to authorized tangible copies made outside the copyright scheme. But whatever might be meant, it is not clear how the "later use" aspect is to be administered, by whom, and under what standards.

D. "Contractual Obligations"

Section 108(f)(4) provides that "[nothing in this section] in any way affects . . . any contractual obligation assumed at any time by the library or archives when it obtained a copy or phonorecord of a work in its collections."[248] Thus the statute, which by making exemptions for certain kinds of photocopying either assumed there would be no economic harm to the copyright owner or as a matter of public policy reallocated the costs for such use, suggests the means to correct imbalances that might in fact have occurred. A periodical, for example, would appear to have the option of leasing rather than selling subscriptions, and thus of either barring photocopying by libraries by contract—presumably by means of a revocable lease—or of licensing photocopying by libraries under the royalty provisions.

In fact, even if the major assumption of the Act and of the libraries with respect to a periodical is correct (namely, that the degree of photocopying permitted by the statute would have no serious effect on copyright-scheme economics), it is difficult to see how a new periodical with a narrow readership (such as a specialized journal) could be launched in any way other than by lease or contract. There would otherwise appear to be no reason for a library to subscribe to the whole periodical, since, for all the library knows, individual articles needed might well be within permissible photocopying limits paid for by the patron; there would appear to be little need for an individual scholar to subscribe, since all he would require would be a table of

---

[248] The Committee Reports are silent on this provision, but the phrase "at any time" clearly refers to contracts antedating the new Copyright Act. It is not clear how far Federal preemption of "legal and equitable rights . . . equivalent to . . . copyright," set out in §301, would bar certain contracts dealing with newly published works.

contents of an issue either for ordering purposes or for making a copy of an article himself; and there would appear to be no way to assure a potential advertiser, whose payments might help support the journal, either that the audience would be of a certain size or, indeed, that anyone would see the advertisement at all.

It is hard to see how the contractual obligation route could be applied to ordinary books, but there would appear to be no bar to applying it to those professional reference works kept current by weekly, monthly, or yearly supplements, and perhaps to other reference works as well, particularly those sold mostly to institutions.

But there is another aspect of the "contractual obligations" provision that poses more complicated questions. The first three factors we have been discussing—the obtainability of a copy in the context of the "availability" test of the statute, the "fair price" test, and the "later use" test—have two characteristics fundamentally different from those involved in the "contractual obligation" factor. All three have to do ultimately with "transactions" between the patron-user, who pays for the reproduced copy, and the copyright owner, but all three are essentially *outside* the copyright scheme, the relationship being determined by transactions expressly exempted by Congress and accordingly governed by legislative rules (and court-determined standards), including the rule that for the most part the copy become the property of the patron-user. The "contractual obligation" provision differs from this for two reasons: first, because it is squarely *within* the copyright design, in that what operates is simply a market-determined economic mechanism wholly within the dynamics of the copyright scheme; and, second, because the transaction has directly to do with a legal relationship between the copyright owner and the library, triggered when the library acquires the work itself in a mode involving privity with the supplier, rather than in a statutorily determined agency relationship between user and library. That relationship between copyright owner and library, indeed, is much like that between an author and a publisher. That is, in depositing his dissertation in his school's library, the student-author, for example, has given the library the right under the statute to reproduce as many copies of his work as there are qualifying libraries,[249] unless he expressly restricts the library by means of an agreement when the dissertation is deposited.

---

[249] Section 108(b), which in the absence of a written agreement to the contrary permits a library or archives to reproduce, in facsimile form, a manuscript deposited with it for distribution to another library "for research use," might be seen as putting the library in the publishing business, so long as it limits its customers to other libraries. This could conceivably affect the pub-

Also, the contractual mechanism could be used to hold a library to stricter standards than the statute requires with respect to the policing of photocopying machines, care in handling copyright notices, number of copies for classroom use, and the like.

E. *"Direct or Indirect Commercial Advantage"*

One of the most puzzling provisions of the library reproduction exemption is the condition that the "no more than one copy"[250] permitted be reproduced or distributed "without any purpose of direct or indirect commercial advantage."[251] This curious phraseology appears to reflect both a public policy that certain "nonprofit" purposes or certain institutions, or both, would excuse what would otherwise be an infringement, and a precision of meaning particular to this provision of the statute. But the distinctions implied and the meaning indicated are far from clear.

What is the standard for an "indirect purpose": For a profitmaking enterprise, does something as indirect, but as purposeful, as "public relations"—i.e., an activity that, while it is not "directly" a part of a commercial enterprise, creates good will for the enterprise—come within the exemption? If it does, what is left of the meaning of "indirect"?

Does the designation of an enterprise as "nonprofit" settle the "commercial advantage" question, so that a university-run bookstore, for example, is within the exception? If not, what is the test?

The House and Senate Committee Reports are in conflict, and the Conference Report begs both questions. With respect to a "profit making" or "commercial" organization, the Senate Committee Report

---

lishing economics of a work that might later on, on substantive grounds, recommend itself to houses—such as university presses—that ordinarily publish works with limited, often largely library, markets. That is, if a manuscript or typescript deposited in a library should be seen by, say, some 500 libraries as warranting their having copies of it for research purposes, section 108(b) would permit the depository library to manufacture and sell that many copies to other libraries, at whatever price it chooses, so long, apparently, as they are manufactured one at a time. (At least, "one at a time" would appear to be what § 108(a)'s "no more than one copy" means, since the rest of § 108 clearly does *not* limit a library to making only one copy of a work. But it is not completely clear what is or is not permissible if requests for photocopies from two or more libraries should be received on the same day.) Nothing in the statute limits the number of copies a library may reproduce under section 108(b).

[250] § 108(a).
[251] § 108(a) (1).

is clearly directed at barring the use of the exception, directly or indirectly, in a way that furthers the commercial effort of the organization:

> "The limitation of section 108 to reproduction and distribution by libraries and archives 'without any purpose of direct or indirect commercial advantage' is intended to preclude a library or archives in a profit-making organization from providing photocopies of copyrighted materials to employees engaged in furtherance of the organization's commercial enterprise, unless such copying qualifies as a fair use, or the organization has obtained the necessary copyright licenses."[252]

The House Committee Report says the opposite, permitting photocopying "even though the copies are furnished to the employees of the organization for use in their work." It then sharply limits the meaning of "direct or indirect commercial advantage" to the *intrinsic transaction* involving the exemption: "the 'advantage' referred to in this clause *must attach* to the immediate commercial motivation behind the reproduction or distribution itself, rather than to the ultimate profit-making motivation behind the enterprise in which the library is located."[253] It is not clear how much is left of "indirect commercial advantage," and when the question was raised in the House-Senate committee conference the test was defined in the Conference Report tautologically as "without any commercial motivation," so long as the "criteria" and "requirements" of the section are met.[254]

With respect to a "nonprofit" organization, the Senate Report is silent, except as to what might be inferred from its sole mention of a "profit-making" organization in the quotation above, while the House Report has several paragraphs in which distinctions appear to be made between "nonprofit" institutions and profit-making institutions. (Section 108 itself does not use the term "nonprofit.") But if the exemption is governed by the House Report's narrow definition of *direct* "commercial advantage," there would appear to be no meaningful distinctions between such institutions or organizations, and, in fact, a careful reading of the House Committee Report on the subject reveals none.[255]

---

[252] Senate Committee Report, 67.
[253] House Committee Report, 75. [Emphasis added.]
[254] Conference Report, 74.
[255] Does the phrase "[no] purpose of direct or indirect commercial advantage" therefore mean nothing at all in the institutional context, apart from its being subsumed completely in the "substitute for a...purchase" provision of section 108(g) (2)? An examination of the history of the phrase suggests

In the end, one must conclude that the conflicting and ambiguous committee glosses would be totally useless as a guide to courts in particular cases, and the reason is not far to seek: what has been lost sight of is the fact that except for the special replacement exemption of subsection 108(c), the reproduction permitted under the statute has *solely* to do with an individual *user's* purpose, for which the library is merely a servicing agent. That is, insofar as the library-exemption glosses deal with the individual user's purpose, they are redundant, for the determination of that purpose is governed by subsections (d)(1) and (e)(1), requiring the photocopy to be made solely for "private study, scholarship, or research." There is nothing in section 108 permitting an institution to reproduce a copyrighted work for its own purposes except to repair a damaged copy or replace a lost one. And insofar as there might be limitations on what institutions might act as a photocopying agent in what circumstances, the silence, and therefore the ambiguity, of the statute is not illuminated by the attempts of the House Committee Report to deal with it. We examine several questions that suggest themselves by this ambiguity.

The language of subsection 108(a)(2) would permit the owner of almost any bona fide collection of works to act as a photocopying agent: whatever the limitations implied by so ambiguous a term as "open to the public," the sole requirement that a collection be "available . . . to other persons doing research in a specialized field" is wide enough, for practical purposes, to leave no institution or collection of works outside it. There are no enforceable limitations on "specialized fields," no encumbrances on the meaning of "available," no requirement that the collection be publicly owned or nonprofit, and accordingly no definition of the terms "library or archives" that

----

that this may very nearly be the case. The phrase first appeared in the bill passed by the House in 1967 [H.R. 2512], where section 108 dealt solely with the reproduction of *unpublished* archival material theretofore protected only by common-law copyright—the sort of facsimile reproduction of manuscripts and other such material now provided for in section 108(b) of the Act. The legislative report of the House Judiciary Committee accompanying the 1967 bill, Report No. 83, Mar. 8, 1967, 90th Cong., 1st Sess., explained that what is meant by no "purpose of direct or indirect commercial advantage" was that "no facsimile copies or phonorecords made under this section can be distributed to scholars or the public; if they leave the institution that reproduced them, they must be deposited. . .in another 'nonprofit institution'." Since what the 1976 Act's sections 108(d) and (e) permit—distribution of photocopies to individual scholars (though of course of *published* works)—is precisely what according to the 1967 House report the phrase "direct or indirect commercial advantage" meant to bar, the phrase insofar as its meaning can be derived from its origin is nullified by the terms of the rest of the new statute.

imposes workable restrictions on the agency itself. A "library" or archives" could be a bona fide collection of works on the Loch Ness monster.

There is accordingly with respect to the test of *availability* of a collection to "persons doing research in a specialized field" no requirement of a nonprofit or "no commercial advantage" standard either for the library itself or for its parent institution. Nor does the test of being "open to the public" require a library to be free. All that the term ordinarily requires is that any member of the general public be admitted on the same terms, whether free or for a fee. In short, the definition of what might constitute a qualifying library, permitted under the copyright law to provide photocopies for individual users, might well be: "almost anything at all."

But that does not end the questions about what agency may actually fabricate a copy of a work, and under what terms.

Since a library may not make a copy of any work for itself, and since the impetus for the exercise of the exemptions set out in subsections (d) and (e) derive solely from the request of an individual user under the requirement that a reproduction be used solely for "private study, scholarship, or research," there would appear to be only two ways in which there could be "commercial advantage" in library photocopying within the criteria of the section, one having to do with individual purpose, the other having to do with the price a photocopying agency charges the individual.

The statutory source for the "private use" test for any photocopying beyond "replacement" of works in a library collection lies in two identically worded parts of subsections (d) and (e) of section 108, which provide that the rights of reproduction and distribution under the exemption apply only if:

> "(1) The copy becomes the property of the user, and the library or archives has had no notice that the copy would be used for any purpose other than private study, scholarship, or research."

The test for user copying, thus obliquely stated in terms of the copying agency's knowledge of the user's purpose, is no more directly expressed in the Committee Reports' comments on these subsections, which merely repeat the words of the statute. But in the context of "direct or indirect commercial advantage," the expansiveness of the terms "study, scholarship, or research" and the elusiveness of the word "private" could raise troublesome questions for courts.

For example, it is not clear what the word "private" syntactically modifies. There can be no question about its modifying the word "study," since it is the first word in the series. But since "scholarship" is by definition something in the service of public purpose—the disinterested determination of facts or of other truths—"private scholarship" is redundant. And if "private" does not modify "scholarship," then the word is not a series modifier and accordingly does not modify "research" either. If that is the case, does the inclusiveness of the notion of "research"—which embraces all intellectual inquiry, without regard to its purpose—cancel the "commercial" purpose test altogether? Analyzed this way, the notion of "private" as applied to research disappears from section 108 altogether, a result consistent with the House Report's comment on the permissibility of the use of photocopies by employees for their employers' purposes.[256]

The wording of the British statute removes the ambiguity in the same context. The photocopying section of the British copyright act uses the phrase "for purposes of research and private study,"[257] thereby seeming to make clear that the word "private" does not modify "research": *all* research would appear legitimately to be within the purposes for which a person may make a photocopy, though a British court could conceivably decide otherwise.

This clarity in the British statute—if, indeed, there is in fact no dispute about the absence of limitations on the meaning of research—derives in part from its making clear the distinction between the photocopying agency and the user, and accordingly reflects a coherent legislative view of the relationship of the photocopying exemption to the internal dynamics of the copyright scheme: the British act expressly requires that the library be nonprofit, and expressly requires that the cost charged to the person to whom a copy is supplied include not only all costs of production, but also overhead costs of the library.[258] The British act would appear to recognize, despite its careful language to the contrary, that no test of "purpose" except for resale of the photocopy can practicably be established by statute, and to be guided by the central principle that the basic risk taken by society in the *economic* efficiency of the copyright scheme not be opened to question by governmental subsidizing of photocopying costs. In short, the British act deals with the technological imperatives of photocopying by accommodating the copyright mechanism to the *access* requirements of photocopying, rather than to a modification of the *cost* factors in the copyright scheme. This observation, which

---

[256] *See* text at n. 251, *supra.*
[257] Copyright Act, 1956, 4 & 5 Eliz. 2, § 7(2) (b).
[258] *Id.* at § 7(2) (e).

comes close to identifying the source of the basic conceptual confusion with the new Copyright Act's approach to library photocopying, brings us to the sixth topic illustrative of the administrability questions raised by the conceptual difficulties in the photocopying exemption.

F. Photocopying Prices: The Open Question

The two transaction-cost questions involved in the mechanism of library photocopying under the section 108 exemption—Who pays for the making of the photocopy for an individual? and, How much is the library to charge?—are not addressed directly by the Copyright Act. The second question is not in any way raised, but the statute's answer to the first question may be obliquely, but hardly definitively, suggested by the requirement in sections (d)(i) and (e)(i) that "the copy becomes the property of the user." That is, the proviso that the copy become the property of the user suggests that the user might be expected to pay for it, but the statute does not *require* that the recipient of the photocopy be charged anything at all. It is unclear what meaning, if any, attaches to the absence of any reference in the statute to photocopying costs and to the requirement that any photocopy made of a journal article (under subsection [d]) or of a book (under subsection [e]) become the property of the user. Since the British act seems unequivocal on both factors,[259] we might find it easier to dis-

---

[259] Though the British statute appears unequivocal, whether or not a library can nevertheless make a copy of an article, of a complete work, or a part of a work for *itself*, as opposed to for a patron, is obscured by the subsequent regulations of the Board of Trade. The statute itself says "that the copies are [to be] supplied only to persons satisfying the librarian. . .that they require them for purposes of research or private study and will not use them for any other purpose," § 7(2) (b), but the Board of Trade regulations implementing the statute, though for the most part also unambiguous, at one place seems to say expressly that libraries *can* obtain photocopies for themselves: "*librarians* to whom copies are supplied shall be required to pay for such copies. . ." Board of Trade Regs. S.I. 1957, No. 868, § 5(b). [Emphasis added.]

It would be inconsistent with what otherwise appears to be the letter and spirit of the British Act if a library could supply itself with a copy of an article or of a work for its collections generally, and though the Board of Trade regulations seem to imply that it *can*, it is clear that the librarian can assure the rest of the requirement, that the use be *solely* "for research and private study," only by not putting it on the shelves of its general collections or not generally lending it without the same kind of controls as required initially for photocopying. That would appear to bar an expansive reading of the Board of Trade regulations—which isn't to say that such a reading by a court could nevertheless not be made.

cern the Congressional conceptualization of the copyright-scheme dynamics in the United States act, which is not so clear, by first examining the British act.

With respect, then, to the primary copyright-scheme elements of control of access and control of costs, nine inferences can be drawn from the requirements in the British act *both* that the photocopying exemption be available only to individual users and that such users pay at least the total costs, direct and indirect, of making the copy:

(1) that the copyright-scheme mechanism giving exclusive control of use to the author would ordinarily bar the sort of access provided by the exemption;

(2) that the making of a copy under the terms of the exemption reallocates some costs—that is, makes an economic adjustment in the scheme (otherwise, there would be no need to restrict the class of consumers);

(3) that the substantive principle on which the exemption is based is the superior public interest in the private purposes of individuals;

(4) that the adjustment in copyright-scheme controls is basically in control of access, rather than in the reallocation of costs;

(5) that with respect to cost the reliance is still on the risk taken by society in the efficiency of the copyright scheme;

(6) that accordingly there must be costs borne by the user who chooses to use the access exemption rather than to purchase the normal copyright-scheme commodity;

(7) that such costs must be at least the true cost of manufacture,

(8) but that such costs are not to include the author's royalty;

(9) a reallocation of which costs will not essentially disturb the author's incentive.

A vital fact reveals itself from an examination of the scheme set out in this list: the exemption does not in fact have to do with the interests of nonprofit institutions (whatever the wording of the statute), something that leads to a tenth inference (10), namely, that nonprofit institutional consumers are a normal part of copyright-scheme economics. Only inferences 6 and 7 do not clearly apply to the United States statute. There might also be some question about inferences 4 and 5.

The failure of the U.S. statute to require the user to pay for the making of a photocopy, let alone to require that what might be termed "true" costs be charged, expressed the uncertain Congressional conceptualization in the U.S. Act already noted in connection with the "direct or indirect commercial advantage" test, namely, the confusion between individual and institutional purpose stemming from the lack of clarity about the role of the library as the photocopying agent of the patron. The absence of the cost requirement in the U.S. statute leaves ambiguous the Congressional view of the relationship of the exemption to copyright-scheme dynamics, for the wide-open option of the library to charge little, or a great deal, or nothing at all introduces a substantive element in the library's role that contradicts the otherwise service nature of its function.

But inferences 6 and 7 raise the two fundamental sets of questions, under both the British and the U.S. statutes, mentioned at the outset of our discussion of price, though more sharply and more disturbingly under ours. These questions, somewhat rephrased, are: (1) How much may be charged the user by the photocopying agency for the manufacture of a photocopy, and what are the consequences of such charges on copyright-scheme dynamics? and (2) What is the institutional effect of the option given certain institutions, particularly educational institutions, to charge or not to charge for making a photocopy for an individual patron?

1. Photocopying Prices.

The British statute sets a minimum price—"not less than the cost (including a contribution to the general expenses of the library) attributable to their production"--but imposes no limits. There is nothing in the statute that bars the inclusion in the photocopying price of library expenses over and above those attributable pro-rata to "general expenses"—for example, expenses for the purchase of books and periodicals—though the Board of Trade could presumably do so by regulation.[260] The silence of the U.S. statute permits a completely open-ended price. So long as the American library is nonprofit (not a requirement of the U.S. statute, though it is of the British), the same economic questions apply as are raised under both statutes: If the rationale for the photocopying exemption derives from the purpose of the individual user, should not the pricing mechanism be "neutral" in the process of supplying him with a photocopy, in the sense that no factors should affect price other than those attributable to the actual costs (whatever those are) of making the photocopy? Ought the exemption from the copyright scheme to permit the use of the transaction, by which the user acquires the tangible expression of an author's work, for purposes having nothing to do with the exemption—such as reallocating costs as between institution and individual?

There is nothing to bar libraries in Great Britain from using commercial photocopying suppliers, since all that is required is that

---

[260] *Id.* at § 7

the work be supplied "on behalf of a privileged library." The United States statute would also appear to permit photocopying by commercial organizations, provided the provisions of the library photocopying exemptions are met. But the fact that the U.S. statute leaves open the possibility of a nonprofit library's charging more than the actual costs of photocopying introduces a new dimension to the photocopying exemption, namely, that "profits" could be made in the sale of the photocopy though the author be barred by statute from participating in the economic transaction.[261] If that is the case, something almost certainly not contemplated by Congress has entered into the legislative adjustment of the copyright scheme.

2. A New Institutional Tension

Whatever the dimension of the library photocopying exemption, the requirement of the statute that the patron own the photocopy requires educational institutions to face a novel, difficult, and important policy decision: Who is to pay for the photocopy when the library does not own the work to be copied? If the library has the work in its collections, there is no matter of principle or of institutional policy at stake: since the work has been made freely available by the institution, the photocopy in this circumstance is solely for the patron's convenience, and there is no reason why he should not pay for it. The customary, library-patron relationship is unchanged. Similarly, if the library simply absorbs the cost of obtaining photocopies for its own

---

[261] Consider, for example, the implications not only of the manufacturing option but of the distribution option as well. Section 108(a) begins by defining the exemption as one permitting a library or archives to reproduce or to distribute a copy of a work—that is, the exemptions are severable. That subsections (d), (e), (f), (g), and (h) describe the double exemption in "and" form does not alter the severability of its parts. Though the "or" formulation means that a library may under the exemption distribute a copy that it has not itself actually manufactured, the strictures on what may be "distributed without any direct or indirect commercial advantage" are not easy to define.

If a university (or a library) decides to license a commercial concessionaire on its campus to run its copycenter may it therefore not have a work copied under the statute, even though the criteria of section 108 are met? And if the answer is that it can, is it different if the copycenter is across the street, on private property? Does the profit such a licensee makes, or the license fee the university receives, "attach" within the prohibited meaning of the House Committee Report? If the copycenter is simply another part of the university, may it not recover more than its "direct" cost? Is it the view of the statute that a nonprofit institution does not, by definition, engage in any profitable activity?

bona fide institutional patrons, the usual relationship between, say, student and university, would be undisturbed. That is, whatever the acquisitions-policy reasons for not having a work in its collections, the absorption of the photocopying costs by the library puts the transaction in the same mode as an ordinary interlibrary loan: the sole burden on the patron in both cases is the unavailability of the book in his own library and a consequent increase in the difficulty and speed of access.

But if the work is not available in its own collections, and if the student or teacher is charged for the photocopy made for the user at a remote library, there is interposed between the user and access to the work a monetary decision about whether or not the work will be consulted at all. In that event a change of some dimensions has been made in the workings, for example, of the academic or public enterprise—in the rule that a work to be consulted would be freely available to the institutional patron whether or not the work were in its own collections or had to be borrowed via interlibrary loan from another library. The "public good" aspect of a library's collections, whereby once acquired a work is freely available to all in the library's community, will have been diluted.[262] And insofar as an institutional decision to acquire or not to acquire a particular work reflects a policy with respect to access—that is, with where the work is to be located, rather than with whether the work is likely to be consulted—there is a transfer of library costs directly to a particular user in the degree to which he or she consults such works.

It is unclear whether or not the real costs of making photocopies and servicing orders for them in general equals or exceeds the costs of normal acquisitions,[263] but insofar as the transfer of costs, for whatever reason, from libraries to individuals, is to influence individual decisions about either buying a work or making a copy of it, or a part of it, the British act reflects the legislative view that the real costs involved be assessed on the user. The U.S. statute by its silence on photocopying charges poses a complex challenge to American institutions that the British statute preempts for theirs.

What can in general be said about the library exemption for photocopying is that the two tests of *purpose* that the statute attempts to use to limit photocopying—the "without commercial advantage" test of the libraries, and the "private use" test of the individual—turn out on examination to be without real substance. The incantatory "without

---

[262] With respect to the fee-for-service question, *see Double Taxation,* 101 Lɪʙʀᴀʀʏ Jᴏᴜʀɴᴀʟ, No. 20, Nov. 15, 1976, p. 2321.

[263] *See* n. 224, *supra.*

commercial advantage" library test seems not really to have any consistent theoretical base unless all "nonprofit" activities are seen to be outside copyright-scheme commodity economics, as they are not for any other kind of market economics, and insofar as it is attempted to be applied anyway, it is full of loopholes, deliberate or inadvertent. Similarly, the "private study, scholarship, and research" test under section 108 has been brought into question by the House Committee Commentary. In addition, since the phrase includes two terms—"scholarship" and "research"—identical with the "fair use" examples given in section 107 and embodying the notion of "private study," it is hard to see how.a particular user is to know the difference, if there is one, or, as a practical matter, to abide by whatever difference he thinks he sees.

Where this might lead, with respect both to actual practice under the Act and to the formulation of an administrable scheme accommodative of reasonably unhampered photocopying, we suggest in the course of considering, as we do in Chapter V, the "effect" test put into the act by the addition of the "provided" phrase in section 108(g)(2), and, in the light of it, what might constitute a coherent legislative approach to the exclusive-rights tensions in copyright when the scheme is generally under pressure from very great technological or institutional change.

V.

A Functional Approach
to Statutory Adjustments
in the Copyright Scheme

Our examination of how the new Copyright Act deals with the fair use notion and with particular exemptions prepares us to address the general questions posed in Chapter I, namely, What do these changes tell us about the general configuration of copyright under the new Act? and, What does the way in which Congress has dealt with particular exclusive-rights tensions in copyright tell us about both the solution to old problems and an appropriate approach to new ones? That this is not a futile academic exercise in finding unnecessarily precise general principles in a pragmatic mechanism is perhaps suggested by the requirement in the Act[264] that the library photocopying solution arrived at by Congress be reexamined in five years to see how it has worked.

A coherent approach to appropriate statutory adjustments in the copyright scheme requires a clear understanding not only of what its theoretical basis generally consists in—something we examined in Chapter I—but also of how, in fact, a particular law has worked in practice. For if copyright depends essentially on its internal dynamics, as we have argued it does, adjustments in the scheme ought to be limited with some precision to correcting imbalances that the scheme itself requires.

What must be observed at once—and what is frequently seen to be otherwise[265]—is that the 1909 Copyright Act, like its forebears, was in its essentials a simple mechanism, and that it was remarkably resilient in accommodating technological and economic change. It managed reasonably over the years to accommodate, by court decisions

---

[264] § 108(i).

[265] *See, e.g.,* the discussion in the text at n. 47, *supra,* of the characterization of the 1909 Act by the Register of Copyrights in his 1961 Report to Congress.

and legislative amendments, the enlargement of its scope to include new forms of creative work as well as new uses of protected works. And litany-like complaints to the contrary notwithstanding, the workings of the scheme itself permitted voluntary mechanisms (like the performing-rights societies' licensing scheme for the performance of music) that dealt appropriately with problems of access and cost raised by new technologies or by new means of economic exploitation of copyrighted materials.

The old Act could, indeed, have harbored most of the essential accommodations in the new Act: the effects of compulsory licensing systems of the new Act could equally have been accomplished by voluntary collective licenses; there is nothing in the "private agreement solution" to school photocopying that was not possible under the old Act; and it is perhaps not too much to say that even the troublesome question of ubiquitous photocopying could have found solution through acceptable contractual means under the old Act, given enough time for the system to respond to the many forces released by that phenomenon.

But we do not mean to push this too far. The purpose here is to make the point that if it is fundamentally to be relied upon, the copyright scheme is essentially self-policing with the aid of courts, and that adjustments in its basic mechanism ought appropriately to be designed to allow it to work, not to make substantive extra-copyright balancing changes that reconsider and readjust the basic risks.

As we have seen, the new Copyright Act essentially responds to that notion. The reallocation of costs for public policy reasons have for the most part been few and minor. The major failure of the new Act would appear to be with photocopying, to which we must again turn—but with, it is to be hoped, a clearer sense of what is required, and why. As it happens, there is to be found a clue in an interesting recent phenomenon in copyright countries that for some fifteen or twenty years has gradually taken hold in countries other than the United States, namely, the notion of the "public lending right."

A. Accommodating a Governmentally Introduced Imbalance: The Public Lending Right Notion

The idea of the public lending right is that there be payment to an author when his work is borrowed from a public library as well as when it is purchased in the marketplace. Such a right was by the end of 1976 already in effect, either by statute, administrative decree, or collective agreement, in a number of countries, notably Sweden, Denmark, the Netherlands, Australia, and New Zealand, and in West

Germany as part of the copyright act, though payments had not yet begun in the last, and it has been making its way slowly through the British parliamentary system since 1960, when a bill[266] on the matter was introduced. The public lending right scheme generally provides that payment will be made in accordance with the degree of use (both by means of loans to library patrons and by means of use of the library) made of an author's work by public libraries.[267]

It reflects a sense that the author's appropriate expectations of reward for his work are not properly taken care of by the ordinary economics of the copyright scheme, that somehow the nature of public libraries themselves—at least as they have come to be developed in some countries—unacceptably dilute those expectations. For example, during the debate on the Public Lending Right bill in the House of Commons, in 1973, figures were quoted to show that in 1971 in Holland 18 books were borrowed for every 12 bought, in the United States 13 for every 14, and in the United Kingdom 38 for every 4.[268] The same point was made at the second reading of the bill in the House of Lords in March 1976:

> "In 1920 for every book we borrowed from the public lending library we bought ten books in a bookshop. Today it is safe to say that that ratio has been reversed and that for every book we buy we borrow ten."[269]

---

[266] Bill to amend the Public Libraries Act of 1892, 9 Eliz. 2, Bill No. 35.

[267] There are also in the notion of the public lending right aspects of the sort of author interests represented by performance rights in a work. That is, public lending right has to do with a later use of a copyrighted work in which the author continues to have an interest after the essential transaction with respect to the tangible work and involving "economic reward" has taken place. As with those intangible performance interests, public lending right is something recognized by society as "appropriately within the expectations of the author."

[268] Quoted in THE BOOKSELLER, May 12, 1973, at 2405.

[269] Quoted in THE BOOKSELLER, April 10, 1976, at 1920. In the House of Commons debate, Mr. St. John-Stevas, M.P., put the matter this way:
"[T]he novelist, for example, may sell 3,000 copies of a first or subsequent novel which can be borrowed about 117,000 times from libraries—that is the average calculation. The author will obtain between £600 and £800 for the sale of the book but for the 100,000 borrowings he will receive absolutely nothing. The situation is pretty acute in Britain because of the excellence of our library service...They average 600 million loans a year...In the United States, with a population four times that of Great Britain, the circulation from libraries is 450 million loans a year. It is excellent that we should have such a good library service, but it is right that compensation should be paid to the authors who, above all, make the service possible."
Quoted in THE BOOKSELLER, June 5, 1976, at 2529.

Now the principal question about the movement toward the public lending right for our purposes is: "What is going on here with respect to the economics of the copyright scheme?" The following inferences are discernible:

1. that something is wrong with the way the copyright scheme works;

2. that that something is of recent origin;

3. that it has to do with an activity, the functioning of libraries, that has heretofore not only posed no threat to the workings of the copyright scheme, but in fact has been a normal and integral part of the copyright-scheme market;

4. that society's reliance must still be in the essential efficiency of the copyright scheme, in whose viability everyone's interests lie;

5. that the trouble is caused by an activity of government—the establishment, funding, and staffing of public libraries—in a degree so great that the micro-economic balance of the copyright scheme rewards is interfered with;

6. that therefore a statutory solution is needed to correct the imbalance.

The central inference for our purposes is a seventh—namely, that since the reallocation of costs in the copyright-scheme mechanism has been made by a societal policy (funding of ubiquitous lending libraries) that alters the workings of the copyright scheme, those costs must be borne by the central government, and not by either the libraries or the users, whose respective roles as purchasers in copyright-scheme economics would continue unchanged. In short, that the adjustment must be made *outside* the copyright scheme.

The choice of means to correct the maladjustment is instructive: a solution that simply rewarded the author at the user's expense would satisfy the "appropriately expected economic reward" test, but it would build into the copyright scheme an unsettling factor. That is, there would be market-demand decisions—governmental policies about libraries in a dimension that the economics of copyright has not been able to deal with adequately—essentially outside the publishing and marketing mechanism. This "externality" in the market economics of copyright is apparently seen, in this conceptualization, as more than the self-policing copyright mechanism can cope with. At least, that is a conclusion strongly suggested by the growing acceptance in copyright countries both of the notion of the public lending right and of the desirability of funding it from central government sources.[270]

---

[270] This conceptualization is not altered by the fact that in some countries the public lending right may be part of the copyright statute, or that the 1973

In short, the legislative solution involved in public lending right derives from two basic decisions—to reallocate certain costs with respect to the use of copyright materials, and to make the adjustment outside the copyright scheme. The effect is to further both public policies—to preserve the integrity of the copyright scheme and to permit the unencumbered development of public libraries. What light this solution throws on how a legislature might approach another set of public policy issues involving access to and cost of works of the intellect we consider next in a context in which the library again plays a complex agency role, namely photocopying.

B. Accommodating a Technologically Impelled Imbalance: The Three Faces of the Photocopying Problem

The remarkable characteristic of photocopying that makes possible ubiquitous and instantaneous single and multiple reproduction of a work of the intellect is incompatible with one of the fundamental functional bases of copyright, namely, that the author can as a practical matter exert exclusive control of the making and distributing of copies in all significant circumstances. This fact, by changing the normal expectations of society, is recognized by Congress in the Act as requiring a major relinquishment of author control of access. The question posed is as usual dual: In a particular case, what control of *access*, if any, must in the public interest be taken from the author? and, Is the *cost* to be reallocated, or is it to be accommodated within the copyright scheme?

Since the congressional point of departure in this as in other parts of the Act is that the general efficiency of copyright-scheme dynamics is nevertheless to be continued to be relied upon, the

---

bill introduced in Parliament was an amendment to the copyright act, while the 1976 bill was separate from the copyright act. Whether or not the adjustment is statutorily part of the copyright act has partly to do with whether the country concerned is to apply the public lending right not only to its own authors, but also to foreign authors, which adherence to the Universal Copyright Convention would perhaps require if the public lending right were part of a signatory country's copyright act.

As of the end of 1976, only in West Germany was public lending right incorporated in the copyright act. Other reasons given for the change in approach to public lending right legislation in England were that to legislate under copyright would have given writers the right to bargain about their fees, rather than to have them allocated by the government, and that none of the five countries then paying public lending right fees in 1976 paid them to foreign authors. *See* THE BOOKSELLER, May 22, 1976, at 2373.

appropriate approach to these questions, and the success of a particular solution, will depend on the clarity with which both a particular problem is seen and the effect of a particular solution recognized.

However unclearly Congress may have defined 'them as discrete, three distinct problems brought about by photocopying capabilities are dealt with in the Act, each of them having to do with a different element in the functioning of the copyright scheme, all of them deriving from a change required in a fundamental element in the copyright design, namely, that the author relinquish some control of the making of such copies—in short, over access. The three problems, in an order increasingly integral to the functioning of the copyright scheme, have to do with:

(1) the accommodation of a competing public policy of nearly constitutional dimensions, namely, the spontaneous requirements of education in a classroom context;

(2) the accommodation of the technical workings of part of the copyright-scheme distribution mechanism, namely the functioning of libraries and library systems; and

(3) the accommodation of the ultimate requirements of the individual user, whose capability of making copies for himself constitutes what is exactly coextensive with the market outcome of the copyright scheme itself, namely, the user's acquiring a copy of the copyrighted work.

As we have seen, there are only two ways in which exceptions from copyright exclusive-rights controls may be made by Congress: either by an exemption with respect to access alone, by means of a compulsory license with appropriate fees; or by a particularized exemption from all cost and all control of access, with a consequent reallocation of costs outside the copyright scheme. The congressional resolutions of the first and third of the matters set out above, however uncertain the statutory tests and standards and therefore however uncertain the effect in practice, seem reasonably discernible. For the school-use case the Act relies essentially on the dynamics of the copyright scheme itself—i.e., on the workings of marketplace economics, in either a commodity or a contractual mode—but with a limited and narrow exception based on the *spontaneous* needs of teachers. For the personal-use case, the Act provides for a limited exemption for "private" use governed partly by fair use standards and partly by statutory limits so long as "libraries or archives" do the copying.

Though we shall return to these two matters when we consider what approach might encompass all three cases in a coherent and administrable way, our concern first is to arrive at a more precise definition of the nature of the second case, the making of copies in a library context.

1. Interlibrary Transactions Involving Photocopying.

Notwithstanding the language of the statute, which may or may not beg some essential questions, we must begin with the fundamental risk taken in the copyright scheme, and, without characterizing them as good or bad, identify the elements in copyright-scheme dynamics that operate in a particular case. With respect to libraries, then, the three essential facts are: (1) that libraries, as agency arrangements, are substitutes for individual purchases, (2) that accordingly interlibrary transactions, of whatever sort, are in turn substitutes in the first instance for library purchases and in the second for individual purchases, and (3) that in the dimension and with the characteristics these two factors have traditionally had, they have been seen as at the very least accommodatable in the copyright scheme if not, indeed, wholly compatible with it. That is, they have not appeared to disturb the essential workings of copyright-scheme economics.

As we have seen in the case of the public lending right, it is possible for the *magnitude* of what might in economic terms be called an externality—in that instance the application of government funds to the "agency" arrangement—to disturb that economic balance. The two questions posed by the library-growth situation in European countries have been answered, as we have seen, by public policy support of both systems: libraries would continue to be developed and subsidized from central government funds, and insofar as copyright-scheme economics was disturbed, the accommodation was to be made outside the scheme itself, again from central government funds.

The parallel with true interlibrary loans is direct: if, as with subsidized library systems, a program of government subsidization of interlibrary loans were of a magnitude to disturb the normal economic expectations of authors, and if reliance were nevertheless to continue to be on the efficiency of the copyright scheme, the furtherance of both public policies would require an extra-copyright adjustment of economic rewards.

This brings us to photocopying in an interlibrary context: if what photocopying does is to enormously facilitate interlibrary transactions that involve the actual making of new copies, and if the effect is such as to disturb in any degree either an author's expectations or the economic balances in the copyright scheme, a governmental decision of the same character as with public lending right or with the hypothetical government-subsidized interlibrary loan program is required, namely, whether the cost to the copyright scheme is to be dealt with *outside* the scheme—i.e., either defrayed by central government funds or exempted altogether—or *within* the copyright scheme—i.e., by user fees either in a voluntary contractual mode or by means of a compulsory license.

In short, in the first instance the externality that affects the functioning of the copyright scheme is a government allocation of central resources, while in the second it is a technological factor that does so. The ultimate question is the same: Is the cost to copyright-scheme economics to be accommodated by the scheme itself, which means increasing the cost of interlibrary transactions and thereby affecting the development of such a system; or are the costs to be reallocated so as to allow the interlibrary transactions to develop without internal costs?

Unless the dimension of the interlibrary transaction system is simply *defined* as undisturbing of copyright economics, insofar as there is no compensation for the author there has been a reallocation of costs, with the accommodation consequently outside the copyright scheme—as with the public lending right matter, but with the difference that in public lending right the government picks up the external costs, and in the photocopying mode the author does.

The matter is determined in the new Act by decree: it accepts interlibrary loans in the old dimension as undisturbing of copyright-scheme economics and *defines* the statutorily permitted degree of photocopying, insofar as it has to do with libraries rather than with patron-purchasers of photocopies, as leaving that configuration unchanged. Such a definition implies, however, that in an appropriate dimension, interlibrary transactions involving photocopying *would* be disturbing.

The Copyright Act's treatment of library photocopying in an interlibrary transaction mode, then, rests on three premises: the identification of the making of a copy as no different in copyright-scheme economics from the lending of a copy; the presumption that the degree of interlibrary transactions made possible by photocopying and permitted by the Act does not affect copyright-scheme economics; and the conclusion that the limitations in the Act are enforceable.

It is clear, then, that, whatever the degree of confidence with which the first two premises are accepted, the success of the solution depends on the third. An examination of that question brings us to the nexus with the other two photocopying cases—school use and private use—and accordingly to a consideration of what approach might embrace them all in a coherent and administrable scheme.

2. The Compelled Alternatives for the Accommodation of Photocopying in the Copyright Scheme.

The three major photocopying problems dealt with in the Act have all been resolved in the same way, by means of exceptions to the author's exclusive-rights control of the making and selling of copies, characterized in particular cases either as fair use exceptions or as limited exemptions whose economic effects were defined as undisturbing of copyright-scheme dynamics. The obverse of these definitions of limits is that if they are in practice exceeded, the functioning of the copyright scheme would be disturbed. The nexus of all three cases, then, is the likelihood of the limits being observed in practice—in short, whether the scheme will function.

And the thing that all three have in common is the unlikelihood that the relied-upon distinctions can be made, or, if they can, that they would be made, and, in either case, that the scheme is administrable.

As we have seen, none of the distinctions made in the statute for any of the three uses can be relied upon: the "purpose" tests for both individuals and copying agencies are unclear; the guidelines for individuals to make fair-use distinctions imprecise; the distinctions among "noncommercial" libraries, other libraries, and other copying agencies untenable; the noninstitutional controls nonexistent; the responsibility for policing impermissible uses diffuse; the sanction mechanism incompatible with the transient nature of the copying; the administrative burden on the courts too broad.

The awkwardness of the suggested guidelines for library photocopying, together with the limiting of the institutional standards of responsibility for what individuals may do to the mere posting of notices near copying machines, would appear to guarantee that the statutory rules limiting photocopying in libraries will in fact be generally ignored or misunderstood.[271] It is unreasonable to expect a user to dis-

---

[271] For example, the journal AMERICAN LIBRARIES answers a query from a college library about whether or not it is legal under the new Act to microfilm journals that do not themselves supply or arrange for commercial microfilm editions as follows: "A high official at the Copyright Office advises that no one in Washington or anywhere else is qualified to give you a definitive yes or no answer. However, the expert did suggest that Section 108(b) and (e)...would seem to permit reproduction." Vol. 8, No. 2 (February 1977), at 68. The interpretation of the advice, if it was in fact given, would appear to be wholly wrong: section 108(b) deals only with unpublished works, and section 108(e) requires that the copy become the property of the patron-user. The article is silent on whether the publisher of the journal was asked for permission to microfilm the edition, with or without fee.

tinguish between what is fair use, what is exempted use, and what is neither. And when Congress itself has failed to do so with a precision in which it has much confidence, courts will not find it easy to bring a photocopying excess about which there is any doubt at all within the reach of the enforcement provisions of the liability clause, section 504. As our examination of the educational exemption has brought out, the same questions apply there with respect to the ability, interest, and concern of individuals to make the appropriate distinctions.

But perhaps the most persuasive evidence of the defectiveness of the Copyright Act's approach to photocopying is its utter silence about the making of copies outside of institutions, as if the ubiquitous presence of photocopying machines in offices, homes, and commercial service agencies did not constitute something different in the way of danger for the observance of legal constraints—did not render the setting of limits in section 107 and 108 illusory. Compared even with the analogously difficult matter of home taping of sound television,[272] the likelihood of widespread violation of copyright by photocopying is surely in a class by itself.

In these circumstances the self-policing dynamics of copyright are in danger, and accordingly so are public policies that rely on them.

Faced with a similar problem in connection with cable television, Congress in the end had first to decide whether to continue to preserve the integrity of the copyright scheme, and then to make a decision involving an allocation of costs. The problem was this: there were secondary transmissions carried by commercial cable systems that in terms of public policy ought not to trigger copyright controls, but any mechanism to try to distinguish these from the rest of the operation of cable systems involved the making of distinctions, the development of administrative complexities, and policing difficulties that either could not be relied upon or that would cost more than the benefits. The solution was accordingly a simple one, involving two decisions. The first was a technological matter: since no distinction among the uses could practicably be made, none would be. That left only a question of public policy: should all such uses be free—that is, should the accommodation be made *outside* the copyright scheme by means of a reallocation of costs? Or should the extra costs—those involving pay-

---

[272] The complex issues involved have been joined in a suit filed in November, 1976, in federal district court in Los Angeles by two motion-picture companies, Universal Films and Disney Enterprises, owners of copyrighted films, against Sony Corporation of America, manufacturers of home videotaping machines, and their advertising agency, several dealers, and a private user.

ment for what would otherwise be free use—be accommodated *within* the copyright scheme? Either decision would be both simple and administrable. Congress chose the second on public policy grounds—that it was necessary for the functioning of the copyright scheme, and that it was in the public interest to rely on the scheme's internal efficiency.

The same series of questions, it would appear, are posed with respect to photocopying once it is conceded that as a practical matter the distinction between permitted and protected uses either could not or would not be made. In this circumstance, Congress clearly has only two options consistent with permitting the freest possible access to copyrighted work by means of photocopying: either greatly to expand the exemption for photocopying so as to leave a narrow area protected by the author's exclusive-rights control under copyright, or to in effect require an accounting, by compulsory or voluntary licensing, of all photocopying of copyrighted works permitted by the statute. The former solution would reflect the judgment that there is little risk that the economic efficiency of the copyright scheme would be disturbed; or, if it is, that a major reallocation of costs, essentially outside the copyright scheme, is appropriate. The latter solution would reflect a judgment that the essential reliance is to be on the efficiency of the copyright scheme, and that the charge for what would otherwise be a cost-free "fair use" or exempted use is an appropriate societal cost of accommodating both public policies.[273]

The basic rationale of the Copyright Act, as well as the particular Congressional treatment of every copyright issue except the making of copies,[274] compels the conclusion that the only coherent statutory solu-

---

[273] The congressional committees were aware, through the strong participation of the Copyright Office, of the enormous and intelligent effort made in a series of international conferences on the photocopying question, in which the entire spectrum of approaches to solutions to photocopying had been exhaustively gone into. *See* R. E. BARKER, PHOTOCOPYING PRACTICES IN THE UNITED KINGDOM (London, 1970); the 1968 and 1975 reports of UNESCO's Sub-Committee of the Intergovernmental Copyright Committee on Reprographic Reproduction of Works Protected by Copyright, in COPYRIGHT BULLETIN, Vol. II, No. 3, 1968, and Vol. IX, No. 2/3, 1975; and the report of UNESCO's Working Group on Reprographic Reproduction of Works Protected by Copyright, Paris, 2-4 May, 1973.

The discussions and debates at these international conferences are instructive. What begins as simply stated "interest-group" conceptualization gives way gradually not only to the laying out of the complex tensions in the notion of copyright, but also to an awareness that the public interest is somehow bound up with the internal economic dynamics of the copyright scheme. Thus, the 1975 meeting of the UNESCO subcommittee in Washington, D.C., in June, 1975, concluded its deliberations with a resolution to the photocopying problem is as complete an accountability as possible of all such copying.[275] Such a solution does not mean that all photocopying would incur copyright costs: part of the aggregate copying could be made cost free whether access is accomplished in a compulsory licensing mode or a voluntary licensing mode.[276] But what it does mean is that the essential requirement of easy access to copies of copyrighted works posed by the enormous force of ubiquitous photocopying would be met, and that the internal logic of the copyright scheme would be undisturbed.

What is clear is that when Congress after five years of experience under the new Act takes a look at how sections 107 and 108 have

---

tion that, while uneasily straddling the issue, points toward a comprehensive licensing scheme of some sort.

Section 2 of the resolution recommended that "in those States where the use of processes of reprographic reproduction is widespread, such States could consider, among other measures, encouraging the establishment of collective systems to exercise and administer the right to remuneration." 9 COPYRIGHT BULLETIN, No. 2/3, 1975, at 44.

[274] §107 and § 108 deal with both sight and sound ("copy or phonorecord"), but the wording and thrust of the commentary in the Committee Reports have almost wholly to do with printed material, the copyright complexities of which we have been likewise nearly wholly concerned with in this book.

[275] Various statutory licensing schemes covering all photocopying have from time to time been proposed—e.g., by Irwin Karp, Counsel to the Author's League of America, Inc., in A "Statutory Licensing" System for the Limited Copying of Copyrighted Works, 12 Bull. Cr. Soc. 197, Item 160 (1965); by Professor Nimmer to the Committee of Experts, UNESCO, 2 COPYRIGHT BULLETIN, No. 3, 1968 [in NIMMER ON COPYRIGHT, at 656.5, the author suggests the possibility of a "judicially created compulsory license"]; and by the German book trade in 1955. See BARKER, supra n. 273, at 72–73. The Register of Copyrights, in her Second Supplementary Report (October–December 1975), suggests the need for a voluntary blanket-licensing scheme: "Beyond this immediate legislative solution, however, there is a fact that must be faced. Right now, there are activities connected with teaching that constitute infringement, not fair use, and these are bound to increase. Everyone seems to assume that they will somehow be licensed and that royalties will somehow be paid, but as a practical matter this cannot and will not be done on an individual, item-by-item basis. We are entering an era when blanket licensing and collective payments are essential if the educator is not to be a scofflaw and the author's copyright is not to be a hollow shell. It is not going to be easy, but once the scope of fair use in this field of activity has been clarified by legislative action, immediate efforts to establish workable licensing or clearinghouse arrangements will have to begin." Chapter II, p. 31.

[276] Such a suggestion is made by David Catterns, Legal Research Officer of the Australian Copyright Council, in "The Americans, Baby" By Moorehouse: An Australian Story of Copyright and New Technology—The Thirteenth Annual Jean Geiringer Memorial Lecture on International Copyright Law, 23 BULL. CR. SOC. 213, Item 255 (1976).

worked, more careful consideration should be given not only to clarifying and simplifying section 107, as proposed in Chapter II of this book, but also to dealing with educational copying in a separate, narrowly focused section of the statute. As for section 108, there may be no middle ground between a compulsory licensing scheme and omitting section 108 altogether. Utter silence on library photocopying beyond fair use would at the very least be consistent with both the rationale and the practical workings of the copyright scheme. Furthermore, there is no evidence that a statutory solution is necessary: not only does the only case (*Williams and Williams*) under the old Copyright Act having to do with the matter derive from a faulty judicial definition of fair use, but the actual solutions to copying beyond fair use even under the new Copyright Act consist essentially of nonstatutory arrangements—that is, of the three "agreements" worked out by private parties.

C. The Copyright Scheme and the New Copyright Act: Some Final Thoughts

Does the logic of the alternative solutions of photocopying govern other copyright problems? When technological strains are put on the copyright scheme is the answer with respect to exemptions and exceptions nearly complete controls or none? Is the premise with which we began our inquiry in Chapter I—namely, that the constitutional balancing in the copyright scheme is fundamental—sound? Does it either explain how legislatures have dealt with the strains in the copyright scheme in the past or provide a conceptual framework for legislatures dealing with similar strains in the future?

We have been led in this book to generally affirmative answers to these questions, for as we have seen, the central factor putting pressures on the copyright scheme has turned out to be, time after time, a change that enormously increased *access* to works of the intellect: phonograph records, radio, television, secondary transmission capability, "performances" in general, photocopying, computer storage and retrieval, wide-scale government sponsorship of libraries.

Each time, the central question with respect to photocopying has not been whether such access would be denied, but what the *cost* allocations should be—that is, whether the author should be paid. And once that question has been seen as central, the legislative answer has up to now always been the same—all costs or none. In short, the copyright scheme has not consisted of a great deal of complex internal balancing. The "how much" question, when it has been asked by legislatures, has been extremely narrowly posed, and, when it has, in what might be called a presumptive form: if the exemption urged for reasons of public policy were essentially undisturbing of copyright economics, then it had a good claim to its being granted, and granted completely—performances of music in schools, churches, agricultural fairs, and other "noncommercial" occasions, performances and certain copying for the blind or the deaf. But as we have seen, and as this catalogue shows, such reallocations have been few and minor. Except for photocopying, there is in the new Act no substantial exemption with respect to costs, and accordingly no fundamental reallocation of economic interests in the copyright scheme. And except for the way in which it has dealt with photocopying, Congress has on the whole properly seen the "exclusive-rights" notion as consisting of two separate considerations—cost allocation and control of access—and it has dealt with technological pressures by essentially addressing the second.

The "how much" question has not, in the end, really been asked by Congress on its own terms—that is, by questioning the basic reliance on the copyright scheme itself, though there is evidence that Congress thought it was doing so: at the end of the commentary on section 107 in both Committee Reports, Congress says that "it is the intent of this legislation to provide an appropriate balancing of the rights of creators and the needs of users." By repeating in 1976 this incantatory phrase of the Register of Copyrights in 1961, Congress has ended as it began, failing to see that in this context the phrase lacks precision—that it is the copyright notion *itself* that does this balancing, and that the appropriate aim of a statutory scheme in coherence with it is not to change it in a fundamental way.

We return, thus, to where we began our examination by taking in Chapter I the "acceptance of the basic risk" route. The other route requires a different sort of examination, one dealing empirically with the costs and relationships involved in the microeconomic functioning of copyright commodities in the marketplace.[277] But over a quarter of a millennium after the Statute of Anne, whose simple design still essentially shapes our copyright statute, there is little reason to think that that route would result in an answer to the question "how much reliance ought we to place on the internal workings of the copyright scheme" different from what Congress has appeared to conclude in the new Act: as much as possible.

[277] The "Preliminary Report" of the National Commission on New Technological Uses of Copyrighted Works, October 8, 1976, describes various data-gathering studies on photocopying being made available to the Commission in the Course of the year 1977. The studies deal essentially with the three issues—the extent of library photocopying of serial publications, the economic effect of such photocopying, and the feasibility of transaction-based or blanket licensing schemes—needed to illuminate the policy question of whether, in order to provide appropriate access to copyrighted serial publications, the cost allocation should be inside or outside the copyright scheme.

Appendix A

Copyright Act of 1976, Selected Sections (Public Law 94-553, 94th Cong., October 19, 1976, Title 17 USC)

§ 101. Definitions

As used in this title, the following terms and their variant forms mean the following:

.   .   .   .   .

To "display" a work means to show a copy of it, either directly or by means of a film, slide, television image, or any other device or process or, in the case of a motion picture or other audiovisual work, to show individual images nonsequentially.

.   .   .   .   .

To "perform" a work means to recite, render, play, dance, or act it, either directly or by means of any device or process or, in the case of a motion picture or other audiovisual work, to show its images in any sequence or to make the sounds accompanying it audible.

.   .   .   .   .

To perform or display a work "publicly" means—
(1) to perform or display it at a place open to the public or at any place where a substantial number of persons outside of a normal circle of a family and its social acquaintances is gathered; or
(2) to transmit or otherwise communicate a performance or display of the work to a place specified by clause (1) or to the public, by means of any device or process, whether the members of the public capable of receiving the performance or display receive it in the same place or in separate places and at the same time or at different times.

.   .   .   .   .

§ 102. Subject Matter of Copyright: In General

(a) Copyright protection subsists, in accordance with this title, in original works of authorship fixed in any tangible medium of expression, now known or later developed, from which they can be perceived, reproduced, or otherwise communicated, either directly or with the aid of a machine or device. Works of authorship include the following categories:

(1) literary works;
(2) musical works, including any accompanying words;
(3) dramatic works, including any accompanying music;
(4) pantomimes and choreographic works;
(5) pictorial, graphic, and sculptural works;
(6) motion pictures and other audiovisual works; and
(7) sound recordings.

(b) In no case does copyright protection for an original work of authorship extend to any idea, procedure, process, system, method of operation, concept, principle, or discovery, regardless of the form in which it is described, explained, illustrated, or embodied in such work.

§ 103. Subject Matter of Copyright: Compilations and Derivative Works

(a) The subject matter of copyright as specified by section 102 includes compilations and derivative works, but protection for a work employing preexisting material in which copyright subsists does not extend to any part of the work in which such material has been used unlawfully.

(b) The copyright in a compilation or derivative work extends only to the material contributed by the author of such work, as distinguished from the preexisting material employed in the work, and does not imply any exclusive right in the preexisting material. The copyright in such work is independent of, and does not affect or enlarge the scope, duration, ownership, or subsistence of, any copyright protection in the preexisting material.

§ 106. Exclusive Rights in Copyrighted Works

Subject to sections 107 through 118, the owner of copyright under this title has the exclusive rights to do and to authorize any of the following:

(1) to reproduce the copyrighted work in copies or phonorecords;
(2) to prepare derivative works based upon the copyrighted work;

(3) to distribute copies or phonorecords of the copyrighted work to the public by sale or other transfer of ownership, or by rental, lease, or lending;

(4) in the case of literary, musical, dramatic, and choreographic works, pantomimes, and motion pictures and other audiovisual works, to perform the copyrighted work publicly; and

(5) in the case of literary, musical, dramatic, and choreographic works, pantomimes, and pictorial, graphic, or sculptural works, including the individual images of a motion picture or other audiovisual work, to display the copyrighted work publicly.

§ 107. Limitations on Exclusive Rights: Fair Use

Notwithstanding the provisions of section 106, the fair use of a copyrighted work, including such use by reproduction in copies or phonorecords or by any other means specified by that section, for purposes such as criticism, comment, news reporting, teaching (including multiple copies for classroom use), scholarship, or research, is not an infringement of copyright. In determining whether the use made of a work in any particular case is a fair use the factors to be considered shall include—

(1) the purpose and character of the use, including whether such use is of a commercial nature or is for nonprofit educational purposes;

(2) the nature of the copyrighted work;

(3) the amount and substantiality of the portion used in relation to the copyrighted work as a whole; and

(4) the effect of the use upon the potential market for or value of the copyrighted work.

§ 108. Limitations on Exclusive Rights: Reproduction by Libraries and Archives

(a) Notwithstanding the provisions of section 106, it is not an infringement of copyright for a library or archives, or any of its employees acting within the scope of their employment, to reproduce no more than one copy or phonorecord of a work, or to distribute such copy or phonorecord, under the conditions specified by this section, if—

(1) the reproduction or distribution is made without any purpose of direct or indirect commercial advantage;

(2) the collections of the library or archives are (i) open to the public, or (ii) available not only to researchers affiliated with the library or

archives or with the institution of which it is part, but also to other persons doing research in a specialized field; and

(3) the reproduction or distribution of the work includes a notice of copyright.

(b) The rights of reproduction and distribution under this section apply to a copy or phonorecord of an unpublished work duplicated in facsimile form solely for purposes of preservation and security or for deposit for research use in another library or archives of the type described by clause (2) of subsection (a), if the copy or phonorecord reproduced is currently in the collections of the library or archives.

(c) The right of reproduction under this section applies to a copy or phonorecord of a published work duplicated in facsimile form solely for the purpose of replacement of a copy or phonorecord that is damaged, deteriorating, lost, or stolen, if the library or archives has, after a reasonable effort, determined that an unused replacement cannot be obtained at a fair price.

(d) The rights of reproduction and distribution under this section apply to a copy, made from the collection of a library or archives where the user makes his or her request or from that of another library or archives, of no more than one article or other contribution to a copyrighted collection or periodical issue, or to a copy or phonorecord of a small part of any other copyrighted work, if—

(1) the copy or phonorecord becomes the property of the user, and the library or archives has had no notice that the copy or phonorecord would be used for any purpose other than private study, scholarship, or research; and

(2) the library or archives displays prominently, at the place where orders are accepted, and includes on its order form, a warning of copyright in accordance with requirements that the Register of Copyrights shall prescribe by regulation.

(e) The rights of reproduction and distribution under this section apply to the entire work, or to a substantial part of it, made from the collection of a library or archives where the user makes his or her request or from that of another library or archives, if the library or archives has first determined, on the basis of a reasonable investigation, that a copy or phonorecord of the copyrighted work cannot be obtained at a fair price, if—

(1) the copy or phonorecord becomes the property of the user, and the library or archives has had no notice that the copy or phonorecord would be used for any purpose other than private study, scholarship, or research; and

(2) the library or archives displays prominently, at the place where orders are accepted, and includes on its order form, a warning of copy-

right in accordance with requirements that the Register of Copyrights shall prescribe by regulation.

(f) Nothing in this section—

(1) shall be construed to impose liability for copyright infringement upon a library or archives or its employees for the unsupervised use of reproducing equipment located on its premises: *Provided,* That such equipment displays a notice that the making of a copy may be subject to the copyright law;

(2) excuses a person who uses such reproducing equipment or who requests a copy or phonorecord under subsection (d) from liability for copyright infringement for any such act, or for any later use of such copy or phonorecord, if it exceeds fair use as provided by section 107;

(3) shall be construed to limit the reproduction and distribution by lending of a limited number of copies and excerpts by a library or archives of an audiovisual news program, subject to clauses (1), (2), and (3) of subsection (a); or

(4) in any way affects the right of fair use as provided by section 107, or any contractual obligations assumed at any time by the library or archives when it obtained a copy or phonorecord of a work in its collections.

(g) The rights of reproduction and distribution under this section extend to the isolated and unrelated reproduction or distribution of a single copy or phonorecord of the same material on separate occasions, but do not extend to cases where the library or archives, or its employee—

(1) is aware or has substantial reason to believe that it is engaging in the related or concerted reproduction or distribution of multiple copies or phonorecords of the same material, whether made on one occasion or over a period of time, and whether intended for aggregate use by one or more individuals or for separate use by the individual members of a group; or

(2) engages in the systematic reproduction or distribution of single or multiple copies or phonorecords of material described in subsection (d): *Provided,* That nothing in this clause prevents a library or archives from participating in interlibrary arrangements that do not have, as their purpose or effect, that the library or archives receiving such copies or phonorecords for distribution does so in such aggregate quantities as to substitute for a subscription to or purchase of such work.

(h) The rights of reproduction and distribution under this section do not apply to a musical work, a pictorial, graphic or sculptural work, or a motion picture or other audiovisual work other than an audiovisual work dealing with news, except that no such limitation shall apply with respect to rights granted by subsections (b) and (c) or with respect to pictorial or graphic works published as illustrations, diagrams, or similar adjuncts to works of which copies are reproduced or distributed in accordance with subsections (d) and (e).

(i) Five years from the effective date of this Act, and at five-year intervals thereafter, the Register of Copyrights, after consulting with representatives of authors, book and periodical publishers, and other owners of copyrighted materials, and with representatives of library users and librarians, shall submit to the Congress a report setting forth the extent to which this section has achieved the intended statutory balancing of the rights of creators, and the needs of users. The report should also describe any problems that may have arisen, and present legislative or other recommendations, if warranted.

§ 109. Limitations on Exclusive Rights: Effect of Transfer of Particular Copy or Phonorecord

(a) Notwithstanding the provisions of section 106(3), the owner of a particular copy or phonorecord lawfully made under this title, or any person authorized by such owner, is entitled, without the authority of the copyright owner, to sell or otherwise dispose of the possession of that copy or phonorecord.

(b) Notwithstanding the provisions of section 106(5), the owner of a particular copy lawfully made under this title, or any person authorized by such owner, is entitled, without the authority of the copyright owner, to display that copy publicly, either directly or by the projection of no more than one image at a time, to viewers present at the place where the copy is located.

(c) The privileges prescribed by subsections (a) and (b) do not, unless authorized by the copyright owner, extend to any person who has acquired possession of the copy or phonorecord from the copyright owner, by rental, lease, loan, or otherwise, without acquiring ownership of it.

§ 110. Limitations on Exclusive Rights: Exemption of Certain Performances and Displays

Notwithstanding the provisions of section 106, the following are not infringements of copyright:

(1) performance or display of a work by instructors or pupils in the course of face-to-face teaching activities of a nonprofit educational institution, in a classroom or similar place devoted to instruction, unless, in the case of a motion picture or other audiovisual work, the performance,

or the display of individual images, is given by means of a copy that was not lawfully made under this title, and that the person responsible for the performance knew or had reason to believe was not lawfully made;

(2) performance of a nondramatic literary or musical work or display of a work, by or in the course of a transmission, if—

(A) the performance or display is a regular part of the systematic instructional activities of a governmental body or a nonprofit educational institution; and

(B) the performance or display is directly related and of material assistance to the teaching content of the transmission; and

(C) the transmission is made primarily for—

(i) reception in classrooms or similar places normally devoted to instruction, or

(ii) reception by persons to whom the transmission is directed because their disabilities or other special circumstances prevent their attendance in classrooms or similar places normally devoted to instruction, or

(iii) reception by officers or employees of governmental bodies as a part of their official duties or employment;

(3) performance of a nondramatic literary or musical work or of a dramatico-musical work of a religious nature, or display of a work, in the course of services at a place of worship or other religious assembly;

(4) performance of a nondramatic literary or musical work otherwise than in a transmission to the public, without any purpose of direct or indirect commercial advantage and without payment of any fee or other compensation for the performance to any of its performers, promoters, or organizers, if—

(A) there is no direct or indirect admission charge; or

(B) the proceeds, after deducting the reasonable costs of producing the performance, are used exclusively for educational, religious, or charitable purposes and not for private financial gain, except where the copyright owner has served notice of objection to the performance under the following conditions;

(i) the notice shall be in writing and signed by the copyright owner or such owner's duly authorized agent; and

(ii) the notice shall be served on the person responsible for the performance at least seven days before the date of the performance, and shall state the reasons for the objection; and

(iii) the notice shall comply, in form, content, and manner of service, with requirements that the Register of Copyrights shall prescribe by regulation;

(5) communication of a transmission embodying a performance or display of a work by the public reception of the transmission on a single

receiving apparatus of a kind commonly used in private homes, unless—

(A) a direct charge is made to see or hear the transmission; or

(B) the transmission thus received is further transmitted to the public;

(6) performance of a nondramatic musical work by a governmental body or a nonprofit agricultural or horticultural organization, in the course of an annual agricultural or horticultural fair or exhibition conducted by such body or organization; the exemption provided by this clause shall extend to any liability for copyright infringement that would otherwise be imposed on such body or organization, under doctrines of vicarious liability or related infringement, for a performance by a concessionnaire, business establishment, or other person at such fair or exhibition, but shall not excuse any such person from liability for the performance;

(7) performance of a nondramatic musical work by a vending establishment open to the public at large without any direct or indirect admission charge, where the sole purpose of the performance is to promote the retail sale of copies or phonorecords of the work, and the performance is not transmitted beyond the place where the establishment is located and is within the immediate area where the sale is occurring;

(8) performance of a nondramatic literary work, by or in the course of a transmission specifically designed for and primarily directed to blind or other handicapped persons who are unable to read normal printed material as a result of their handicap, or deaf or other handicapped persons who are unable to hear the aural signals accompanying a transmission of visual signals, if the performance is made without any purpose of direct or indirect commercial advantage and its transmission is made through the facilities of: (i) a governmental body; or (ii) a noncommercial educational broadcast station (as defined in section 397 of title 47); or (iii) a radio subcarrier authorization (as defined in 47 CFR 73.293–73.295 and 73.593–73.595); or (iv) a cable system (as defined in section 111(f)).

(9) performance on a single occasion of a dramatic literary work published at least ten years before the date of the performance, by or in the course of a transmission specifically designed for and primarily directed to blind or other handicapped persons who are unable to read normal printed material as a result of their handicap, if the performance is made without any purpose of direct or indirect commercial advantage and its transmission is made through the facilities of a radio subcarrier authorization referred to in clause (8)(iii), *Provided,* That the provisions of this clause shall not be applicable to more than one performance of the same work by the same performers or under the auspices of the same organization.

§ 111. Limitations on Exclusive Rights: Secondary Transmissions

(a) Certain Secondary Transmissions Exempted

The secondary transmission of a primary transmission embodying a performance or display of a work is not an infringement of copyright if—

(1) the secondary transmission is not made by a cable system, and consists entirely of the relaying, by the management of a hotel, apartment house, or similar establishment, of signals transmitted by a broadcast station licensed by the Federal Communications Commission, within the local service area of such station, to the private lodgings of guests or residents of such establishment, and no direct charge is made to see or hear the secondary transmission; or

(2) the secondary transmission is made solely for the purpose and under the conditions specified by clause (2) of section 110; or

(3) the secondary transmission is made by any carrier who has no direct or indirect control over the content or selection of the primary transmission or over the particular recipients of the secondary transmission, and whose activities with respect to the secondary transmission consist solely of providing wires, cables, or other communications channels for the use of others: *Provided,* That the provisions of this clause extend only to the activities of said carrier with respect to secondary transmissions and do not exempt from liability the activities of others with respect to their own primary or secondary transmissions; or

(4) the secondary transmission is not made by a cable system but is made by a governmental body, or other nonprofit organization, without any purpose of direct or indirect commercial advantage, and without charge to the recipients of the secondary transmission other than assessments necessary to defray the actual and reasonable costs of maintaining and operating the secondary transmission service.

(b) Secondary Transmission of Primary Transmission to Controlled Group

Notwithstanding the provisions of subsections (a) and (c), the secondary transmission to the public of a primary transmission embodying a performance or display of a work is actionable as an act of infringement under section 501, and is fully subject to the remedies provided by sections 502 through 506 and 509, if the primary transmission is not made for reception by the public at large but is controlled and limited to reception by particular members of the public: *Provided,* however, That such secondary transmission is not actionable as an act of infringement if—

(1) the primary transmission is made by a broadcast station licensed by the Federal Communications Commission; and

(2) the carriage of the signals comprising the secondary transmission is required under the rules, regulations, or authorizations of the Federal Communications Commission; and

(3) the signal of the primary transmitter is not altered or changed in any way by the secondary transmitter.

(c) Secondary Transmissions by Cable Systems

(1) Subject to the provisions of clauses (2), (3), and (4) of this subsection, secondary transmissions to the public by a cable system of a primary transmission made by a broadcast station licensed by the Federal Communications Commission or by an appropriate governmental authority of Canada or Mexico and embodying a performance or display of a work shall be subject to compulsory licensing upon compliance with the requirements of subsection (d) where the carriage of the signals comprising the secondary transmission is permissible under the rules, regulations, or authorizations of the Federal Communications Commission.

(2) Notwithstanding the provisions of clause (1) of this subsection, the willful or repeated secondary transmission to the public by a cable system of a primary transmission made by a broadcast station licensed by the Federal Communications Commission or by an appropriate governmental authority of Canada or Mexico and embodying a performance or display of a work is actionable as an act of infringement under section 501, and is fully subject to the remedies provided by sections 502 through 506 and 509, in the following cases:

(A) where the carriage of the signals comprising the secondary transmission is not permissible under the rules, regulations, or authorizations of the Federal Communications Commission; or

(B) where the cable system has not recorded the notice specified by subsection (d) and deposited the statement of specified and royalty fee required by subsection (d).

(3) Notwithstanding the provisions of clause (1) of this subsection and subject to the provisions of subsection (e) of this section, the secondary transmission to the public by a cable system of a primary transmission made by a broadcast station licensed by the Federal Communications Commission or by an appropriate governmental authority of Canada or Mexico and embodying a performance or display of a work is actionable as an act of infringement under section 501, and is fully subject to the remedies provided by sections 502 through 506 and sections 509

and 510, if the content of the particular program in which the performance or display is embodied, or any commercial advertising or station announcements transmitted by the primary transmitter during, or immediately before or after, the transmission of such program, is in any way willfully altered by the cable system through changes, deletions, or additions, except for the alteration, deletion, or substitution of commercial advertisements performed by those engaged in television commercial advertising market research: *Provided,* That the research company has obtained the prior consent of the advertiser who has purchased the original commercial advertisement, the television station broadcasting that commercial advertisement, and the cable system performing the secondary transmission: *And provided further,* That such commercial alteration, deletion, or substitution is not performed for the purpose of deriving income from the sale of that commercial time.

(4) Notwithstanding the provisions of clause (1) of this subsection, the secondary transmission to the public by a cable system of a primary transmission made by a broadcast station licensed by an appropriate governmental authority of Canada or Mexico and embodying a performance or display of a work is actionable as an act of infringement under section 501, and is fully subject to the remedies provided by sections 502 through 506 and section 509, if (A) with respect to Canadian signals, the community of the cable system is located more than 150 miles from the United States-Canadian border and is also located south of the forty-second parallel of latitude, or (B) with respect to Mexican signals, the secondary transmission is made by a cable system which received the primary transmission by means other than direct interception of a free radio wave emitted by such broadcast television station, unless prior to April 15, 1976, such cable system was actually carrying, or was specifically authorized to carry, the signal of such foreign station on the system pursuant to the rules, regulations, or authorizations of the Federal Communications Commission.

(d) Compulsory License for Secondary Transmissions by Cable Systems

(1) For any secondary transmission to be subject to compulsory licensing under subsection (c), the cable system shall, at least one month before the date of the commencement of operations of the cable system or within one hundred and eighty days after the enactment of this Act, whichever is later, and thereafter within thirty days after each occasion on which the ownership or control or the signal carriage complement of the cable system changes, record in the Copyright Office a notice including a statement of the identity and address of the person who owns or

operates the secondary transmission service or has power to exercise primary control over it, together with the name and location of the primary transmitter or primary transmitters whose signals are regularly carried by the cable system, and thereafter, from time to time, such further information as the Register of Copyrights, after consultation with the Copyright Royalty Tribunal (if and when the Tribunal has been constituted), shall prescribe by regulation to carry out the purpose of this clause.

(2) A cable system whose secondary transmissions have been subject to compulsory licensing under subsection (c) shall, on a semiannual basis, deposit with the Register of Copyrights, in accordance with requirements that the Register shall, after consultation with the Copyright Royalty Tribunal (if and when the Tribunal has been constituted), prescribe by regulation—

(A) a statement of account, covering the six months next preceding, specifying the number of channels on which the cable system made secondary transmissions to its subscribers, the names and locations of all primary transmitters whose transmissions were further transmitted by the cable system, the total number of subscribers, the gross amounts paid to the cable system for the basic service of providing secondary transmissions of primary broadcast transmitters, and such other data as the Register of Copyrights may, after consultation with the Copyright Royalty Tribunal (if and when the Tribunal has been constituted), from time to time prescribe by regulation. Such statement shall also include a special statement of account covering any nonnetwork television programming that was carried by the cable system in whole or in part beyond the local service area of the primary transmitter, under rules, regulations, or authorizations of the Federal Communications Commission permitting the substitution or addition of signals under certain circumstances, together with logs showing the times, dates, stations, and programs involved in such substituted or added carriage; and

(B) except in the case of a cable system whose royalty is specified in subclause (C) or (D), a total royalty fee for the period covered by the statement, computed on the basis of specified percentages of the gross receipts from subscribers to the cable service during said period for the basic service of providing secondary transmissions of primary broadcast transmitters, as follows:

(i) 0.675 of 1 per centum of such gross receipts for the privilege of further transmitting any nonnetwork programing of a primary transmitter in whole or in part beyond the local service area of such primary transmitter, such amount to be applied against the fee, if any, payable pursuant to paragraphs (ii) through (iv);

(ii) 0.675 of 1 per centum of such gross receipts for the first distant signal equivalent;

(iii) 0.425 of 1 per centum of such gross receipts for each of the second, third, and fourth distant signal equivalents;

(iv) 0.2 of 1 per centum of such gross receipts for the fifth distant signal equivalent and each additional distant signal equivalent thereafter; and

in computing the amounts payable under paragraph (ii) through (iv), above, any fraction of a distant signal equivalent shall be computed at its fractional value and, in the case of any cable system located partly within and partly without the local service area of a primary transmitter, gross receipts shall be limited to those gross receipts derived from subscribers located without the local service area of such primary transmitter; and

(C) if the actual gross receipts paid by subscribers to a cable system for the period covered by the statement for the basic service of providing secondary transmissions of primary broadcast transmitters total $80,000 or less, gross receipts of the cable system for the purpose of this subclause shall be computed by subtracting from such actual gross receipts the amount by which $80,000 exceeds such actual gross receipts, except that in no case shall a cable system's gross receipts be reduced to less than $3,000. The royalty fee payable under this subclause shall be 0.5 of 1 per centum, regardless of the number of distant signal equivalents, if any; and

(D) if the actual gross receipts paid by subscribers to a cable system for the period covered by the statement, for the basic service of providing secondary transmissions of primary broadcast transmitters, are more than $80,000 but less than $160,000, the royalty fee payable under this subclause shall be (i) 0.5 of 1 per centum of any gross receipts up to $80,000; and (ii) 1 per centum of any gross receipts in excess of $80,000 but less than $160,000, regardless of the number of distant signal equivalents, if any.

(3) The Register of Copyrights shall receive all fees deposited under this section and, after deducting the reasonable costs incurred by the Copyright Office under this section, shall deposit the balance in the Treasury of the United States, in such manner as the Secretary of the Treasury directs. All funds held by the Secretary of the Treasury shall be invested in interest-bearing United States securities for later distribution with interest by the Copyright Royalty Tribunal as provided by this title. The Register shall submit to the Copyright Royalty Tribunal, on a semiannual basis, a compilation of all statements of account covering the

relevant six-month period provided by clause (2) of this subsection.

(4) The royalty fees thus deposited shall, in accordance with the procedures provided by clause (5), be distributed to those among the following copyright owners who claim that their works were the subject of secondary transmissions by cable systems during the relevant semiannual period:

(A) any such owner whose work was included in a secondary transmission made by a cable system of a nonnetwork television program in whole or in part beyond the local service area of the primary transmitter; and

(B) any such owner whose work was included in a secondary transmission identified in a special statement of account deposited under clause (2)(A); and

(C) any such owner whose work was included in nonnetwork programing consisting exclusively of aural signals carried by a cable system in whole or in part beyond the local service area of the primary transmitter of such programs.

(5) The royalty fees thus deposited shall be distributed in accordance with the following procedures:

(A) During the month of July in each year, every person claiming to be entitled to compulsory license fees for secondary transmissions shall file a claim with the Copyright Royalty Tribunal, in accordance with requirements that the Tribunal shall prescribe by regulation. Notwithstanding any provisions of the antitrust laws, for purposes of this clause any claimants may agree among themselves as to the proportionate division of compulsory licensing fees among them, may lump their claims together and file them jointly or as a single claim, or may designate a common agent to receive payment on their behalf.

(B) After the first day of August of each year, the Copyright Royalty Tribunal shall determine whether there exists a controversy concerning the distribution of royalty fees. If the Tribunal determines that no such controversy exists, it shall, after deducting its reasonable administrative costs under this section, distribute such fees to the copyright owners entitled, or to their designated agents. If the Tribunal finds the existence of a controversy, it shall, pursuant to chapter 8 of this title, conduct a proceeding to determine the distribution of royalty fees.

(C) During the pendency of any proceeding under this subsection, the Copyright Royalty Tribunal shall withhold from distribution an amount sufficient to satisfy all claims with respect to which a controversy exists, but shall have discretion to proceed to distribute any amounts that are not in controversy.

(e) Nonsimultaneous Secondary Transmissions by Cable Systems

(1) Notwithstanding those provisions of the second paragraph of subsection (f) relating to nonsimultaneous secondary transmissions by a cable system, any such transmissions are actionable as an act of infringement under section 501, and are fully subject to the remedies provided by sections 502 through 506 and sections 509 and 510, unless—

(A) the program on the videotape is transmitted no more than one time to the cable system's subscribers; and

(B) the copyrighted program, episode, or motion picture videotape, including the commercials contained within such program, episode, or picture, is transmitted without deletion or editing; and

(C) an owner or officer of the cable system (i) prevents the duplication of the videotape while in the possession of the system, (ii) prevents unauthorized duplication while in the possession of the facility making the videotape for the system if the system owns or controls the facility, or takes reasonable precautions to prevent such duplication if it does not own or control the facility, (iii) takes adequate precautions to prevent duplication while the tape is being transported, and (iv) subject to clause (2), erases or destroys, or causes the erasure or destruction of, the videotape; and

(D) within forty-five days after the end of each calendar quarter, an owner or officer of the cable system executes an affidavit attesting (i) to the steps and precautions taken to prevent duplication of the videotape, and (ii) subject to clause (2), to the erasure or destruction of all videotapes made or used during such quarter; and

(E) such owner or officer places or causes each such affidavit, and affidavits received pursuant to clause (2)(C), to be placed in a file, open to public inspection, at such system's main office in the community where the transmission is made or in the nearest community where such system maintains an office; and

(F) the nonsimultaneous transmission is one that the cable system would be authorized to transmit under the rules, regulations, and authorizations of the Federal Communications Commission in effect at the time of the nonsimultaneous transmission if the transmission had been made simultaneously, except that this subclause shall not apply to inadvertent or accidental transmissions.

(2) If a cable system transfers to any person a videotape of a program nonsimultaneously transmitted by it, such transfer is actionable as an act of infringement under section 501, and is fully subject to the remedies provided by sections 502 through 506 and 509, except that, pursuant to a written, nonprofit contract providing for the equitable sharing

of the costs of such videotape and its transfer, a videotape nonsimultaneously transmitted by it, in accordance with clause (1), may be transferred by one cable system in Alaska to another system in Alaska, by one cable system in Hawaii permitted to make such nonsimultaneous transmissions to another such cable system in Hawaii, or by one cable system in Guam, the Northern Mariana Islands, or the Trust Territory of the Pacific Islands, to another cable system in any of those three territories, if—

(A) each such contract is available for public inspection in the offices of the cable systems involved, and a copy of such contract is filed, within thirty days after such contract is entered into, with the Copyright Office (which Office shall make each such contract available for public inspection); and

(B) the cable system to which the videotape is transferred complies with clause (1)(A), (B), (C)(i), (iii), and (iv), and (D) through (F); and

(C) such system provides a copy of the affidavit required to be made in accordance with clause (1)(D) to each cable system making a previous nonsimultaneous transmission of the same videotape.

(3) This subsection shall not be construed to supersede the exclusivity protection provisions of any existing agreement, or any such agreement hereafter entered into, between a cable system and a television broadcast station in the area in which the cable system is located, or a network with which such station is affiliated.

(4) As used in this subsection, the term "videotape", and each of its variant forms, means the reproduction of the images and sounds of a program or programs broadcast by a television broadcast station licensed by the Federal Communications Commission, regardless of the nature of the material objects, such as tapes or films, in which the reproduction is embodied.

(f) Definitions

As used in this section, the following terms and their variant forms mean the following:

A "primary transmission" is a transmission made to the public by the transmitting facility whose signals are being received and further transmitted by the secondary transmission service, regardless of where or when the performance or display was first transmitted.

A "secondary transmission" is the further transmitting of a primary transmission simultaneously with the primary transmission, or nonsimultaneously with the primary transmission if by a "cable system" not located in whole or in part within the boundary of the forty-eight contiguous States, Hawaii, or Puerto Rico: *Provided, however,* That a nonsimultaneous further transmission by a cable system located in Hawaii of a primary transmission shall be deemed to be a secondary transmission if the carriage of the television broadcast signal comprising such further transmission is permissible under the rules, regulations, or authorizations of the Federal Communications Commission.

A "cable system" is a facility, located in any State, Territory, Trust Territory, or Possession, that in whole or in part receives signals transmitted or programs broadcast by one or more television broadcast stations licensed by the Federal Communications Commission, and makes secondary transmissions of such signals or programs by wires, cables, or other communications channels to subscribing members of the public who pay for such service. For purposes of determining the royalty fee under subsection (d)(2), two or more cable systems in contiguous communities under common ownership or control or operating from one headend shall be considered as one system.

The "local service area of a primary transmitter", in the case of a television broadcast station, comprises the area in which such station is entitled to insist upon its signal being retransmitted by a cable system pursuant to the rules, regulations, and authorizations of the Federal Communications Commission in effect on April 15, 1976, or in the case of a television broadcast station licensed by an appropriate governmental authority of Canada or Mexico, the area in which it would be entitled to insist upon its signal being retransmitted if it were a television broadcast station subject to such rules, regulations, and authorizations. The "local service area of a primary transmitter", in the case of a radio broadcast station, comprises the primary service area of such station, pursuant to the rules and regulations of the Federal Communications Commission.

A "distant signal equivalent" is the value assigned to the secondary transmission of any nonnetwork television programing carried by a cable system in whole or in part beyond the local service area of the primary transmitter of such programing. It is computed by assigning a value of one to each independent station and a value of one-quarter to each network station and noncommercial educational station for the nonnetwork programing so carried pursuant to the rules, regulations, and authorizations of the Federal Communications Commission. The foregoing values for independent, network, and noncommercial educational stations are subject, however, to the following exceptions and limitations. Where the rules and regulations of the Federal Communications Commission require a cable system to omit the further transmission of a particular program and such rules and regulations also permit the substitution of another program embodying a performance or display of a work in place of the omitted transmission, or where such rules and regulations in effect on the date of enactment of this Act permit a cable system, at its election, to effect such deletion and substitution of a nonlive program or to carry additional programs not transmitted by primary transmitters within whose local service area the cable system is located, no value shall be assigned for the substituted or additional program; where the rules, regulations, or authorizations of the Federal Communications Commission in effect on the date of enactment of this Act permit a cable system, at its election, to omit the further transmission of a particular program and such rules, regulations, or authorizations also permit the substitution of another program embodying a performance or display of a work in place of the omitted transmission, the value assigned for the substituted or additional program shall be, in the case of a live program, the value of one full distant signal equivalent multiplied by a fraction that has as its numerator the number of days in the year in which such substitution occurs and as its denominator the number of days in the year. In the case of a station carried pursuant to the late-night or specialty programing rules of the Federal Communications Commission, or a station carried on a part-time basis where full-time carriage is not possible because the cable system lacks the activated channel capacity to retransmit on a full-time basis all signals which it is authorized to carry, the values for independent, network, and noncommercial educational stations set forth above, as the case may be, shall be multiplied by a fraction which is equal to the ratio of the broadcast hours of such station carried by the cable system to the total broadcast hours of the station.

A "network station" is a television broadcast station that is owned or operated by, or affiliated with, one or more of the television networks in the United States providing nationwide transmissions, and that transmits a substantial part of the programing supplied by such networks for a substantial part of that station's typical broadcast day.

An "independent station" is a commercial television broadcast station other than a network station.

A "noncommercial educational station" is a television station that is a noncommercial educational broadcast station as defined in section 397 of title 47.

§ 112. Limitations on Exclusive Rights: Ephemeral Recordings

(a) Notwithstanding the provisions of section 106, and except in the case of a motion picture or other audiovisual work, it is not an infringement of copyright for a transmitting organization entitled to transmit to the public a performance or display of a work, under a license or transfer of the copyright or under the limitations on exclusive rights in sound recordings specified by section 114(a), to make no more than one copy or phonorecord of a particular transmission program embodying the performance or display, if—

(1) the copy or phonorecord is retained and used solely by the transmitting organization that made it, and no further copies or phonorecords are reproduced from it; and

(2) the copy or phonorecord is used solely for the transmitting organization's own transmissions within its local service area, or for purposes of archival preservation or security; and

(3) unless preserved exclusively for archival purposes, the copy or phonorecord is destroyed within six months from the date the transmission program was first transmitted to the public.

(b) Notwithstanding the provisions of section 106, it is not an infringement of copyright for a governmental body or other nonprofit organization entitled to transmit a performance or display of a work, under section 110(2) or under the limitations on exclusive rights in sound recordings specified by section 114(a), to make no more than thirty copies or phonorecords of a particular transmission program embodying the performance or display, if—

(1) no further copies or phonorecords are reproduced from the copies or phonorecords made under this clause; and

(2) except for one copy or phonorecord that may be preserved exclusively for archival purposes, the copies or phonorecords are destroyed within seven years from the date the transmission program was first transmitted to the public.

(c) Notwithstanding the provisions of section 106, it is not an infringement of copyright for a governmental body or other nonprofit organization to make for distribution no more than one copy or phonorecord, for each transmitting organization specified in clause (2) of this subsection, of a particular transmission program embodying a performance of a nondramatic musical work of a religious nature, or of a sound recording of such a musical work, if—

(1) there is no direct or indirect charge for making or distributing any such copies or phonorecords; and

.(2) none of such copies or phonorecords is used for any performance other than a single transmission to the public by a transmitting organization entitled to transmit to the public a performance of the work under a license or transfer of the copyright; and

(3) except for one copy or phonorecord that may be preserved exclusively for archival purposes, the copies or phonorecords are all destroyed within one year from the date the transmission program was first transmitted to the public.

(d) Notwithstanding the provisions of section 106, it is not an infringement of copyright for a governmental body or other nonprofit organization entitled to transmit a performance of a work under section 110(8) to make no more than ten copies or phonorecords embodying the performance, or to permit the use of any such copy or phonorecord by any governmental body or nonprofit organization entitled to transmit a performance of a work under section 110(8), if—

(1) any such copy or phonorecord is retained and used solely by the organization that made it, or by a governmental body or nonprofit organization entitled to transmit a performance of a work under section 110(8), and no further copies or phonorecords are reproduced from it; and

(2) any such copy or phonorecord is used solely for transmissions authorized under section 110(8), or for purposes of archival preservation or security; and

(3) the governmental body or nonprofit organization permitting any use of any such copy or phonorecord by any governmental body or nonprofit organization under this subsection does not make any charge for such use.

(e) The transmission program embodied in a copy or phonorecord made under this section is not subject to protection as a derivative work under this title except with the express consent of the owners of copyright in the preexisting works employed in the program.

§ 113. Scope of Exclusive Rights in Pictorial, Graphic, and Scuptural Works

(a) Subject to the provisions of subsections (b) and (c) of this section, the exclusive right to reproduce a copyrighted pictorial, graphic, or sculptural work in copies under section 106 includes the right to reproduce the work in or on any kind of article, whether useful or otherwise.

(b) This title does not afford, to the owner of copyright in a work that portrays a useful article as such, any greater or lesser rights with respect to the making, distribution, or display of the useful article so portrayed than those afforded to such works under the law, whether title 17 or the common law or statutes of a State, in effect on December 31, 1977, as held applicable and construed by a court in an action brought under this title.

(c) In the case of a work lawfully reproduced in useful articles that have been offered for sale or other distribution to the public, copyright does not include any right to prevent the making, distribution, or display of pictures or photographs of such articles in connection with advertisements or commentaries related to the distribution or display of such articles, or in connection with news reports.

§ 114. Scope of Exclusive Rights in Sound Recordings

(a) The exclusive rights of the owner of copyright in a sound recording are limited to the rights specified by clauses (1), (2), and (3) of section 106, and do not include any right of performance under section 106(4).

(b) The exclusive right of the owner of copyright in a sound recording under clause (1) of section 106 is limited to the right to duplicate the sound recording in the form of phonorecords, or of copies of motion pictures and other audiovisual works, that directly or indirectly recapture the actual sounds fixed in the recording. The exclusive right of the owner of copyright in a sound recording under clause (2) of section 106 is limited to the right to prepare a derivative work in which the actual sounds fixed in the sound recording are rearranged, remixed, or otherwise altered in sequence or quality. The exclusive rights of the owner of copyright in a sound recording under clauses (1) and (2) of section 106 do not extend to the making or duplication of another sound recording that consists entirely of an independent fixation of other sounds, even though such sounds imitate or simulate those in the copyrighted sound recording. The exclusive rights of the owner of copyright in a sound recording under clauses (1), (2), and (3) of section 106 do not apply to sound recordings included in educational television and radio programs (as defined in section 397 of title 47) distributed or transmitted by or through public broadcasting entities (as defined by section 118(g): *Provided,* That copies or phonorecords of said programs are not commercially distributed by or through public broadcasting entities to the general public.

(c) This section does not limit or impair the exclusive right to perform publicly, by means of a phonorecord, any of the works specified by section 106(4).

(d) On January 3, 1978, the Register of Copyrights, after consulting with representatives of owners of copyrighted materials, representatives of the broadcasting, recording, motion picture, entertainment industries, and arts organizations, representatives of organized labor and performers of copyrighted materials, shall submit to the Congress a report setting forth recommendations as to whether this section should be amended to provide for performers and copyright owners of copyrighted material any performance rights in such material. The report should describe the status of such rights in foreign countries, the views of major interested parties, and specific legislative or other recommendations, if any.

§ 115. Scope of Exclusive Rights in Nondramatic Musical Works: Compulsory License for Making and Distributing Phonorecords

In the case of nondramatic musical works, the exclusive rights provided by clauses (1) and (3) of section 106, to make and to distribute phonorecords of such works, are subject to compulsory licensing under the conditions specified by this section.

(a) Availability and Scope of Compulsory License

(1) When phonorecords of a nondramatic musical work have been distributed to the public in the United States under the authority of the copyright owner, any other person may, by complying with the provisions of this section, obtain a compulsory license to make and distribute phonorecords of the work. A person may obtain a compulsory license only if his or her primary purpose in making phonorecords is to distribute them to the public for private use. A person may not obtain a compulsory license for use of the work in the making of phonorecords duplicating a sound recording fixed by another, unless: (i) such sound recording was fixed lawfully; and (ii) the making of the phonorecords was authorized by the owner of copyright in the sound recording or, if the sound recording was fixed before February 15, 1972, by any person who fixed the sound recording pursuant to an express license from the owner of the copyright in the musical work or pursuant to a valid compulsory license for use of such work in a sound recording.

(2) A compulsory license includes the privilege of making a musical arrangement of the work to the extent necessary to conform it to the style or manner of interpretation of the performance involved, but the arrangement shall not change the basic melody or fundamental character of the work, and shall not be subject to protection as a derivative work under this title, except with the express consent of the copyright owner.

(b) Notice of Intention to Obtain Compulsory License

(1) Any person who wishes to obtain a compulsory license under this section shall, before or within thirty days after making, and before distributing any phonorecords of the work, serve notice of intention to do so on the copyright owner. If the registration or other public records of the Copyright Office do not identify the copyright owner and include

an address at which notice can be served, it shall be sufficient to file the notice of intention in the Copyright Office. The notice shall comply, in form, content, and manner of service, with requirements that the Register of Copyrights shall prescribe by regulation.

(2) Failure to serve or file the notice required by clause (1) forecloses the possibility of a compulsory license and, in the absence of a negotiated license, renders the making and distribution of phonorecords actionable as acts of infringement under section 501 and fully subject to the remedies provided by sections 502 through 506 and 509.

(c) Royalty Payable under Compulsory License

(1) To be entitled to receive royalties under a compulsory license, the copyright owner must be identified in the registration or other public records of the Copyright Office. The owner is entitled to royalties for phonorecords made and distributed after being so identified, but is not entitled to recover for any phonorecords previously made and distributed.

(2) Except as provided by clause (1), the royalty under a compulsory license shall be payable for every phonorecord made and distributed in accordance with the license. For this purpose, a phonorecord is considered "distributed" if the person exercising the compulsory license has voluntarily and permanently parted with its possession. With respect to each work embodied in the phonorecord, the royalty shall be either two and three-fourths cents, or one-half of one cent per minute of playing time or fraction thereof, whichever amount is larger.

(3) Royalty payments shall be made on or before the twentieth day of each month and shall include all royalties for the month next preceding. Each monthly payment shall be made under oath and shall comply with requirements that the Register of Copyrights shall prescribe by regulation. The Register shall also prescribe regulations under which detailed cumulative annual statements of account, certified by a certified public accountant, shall be filed for every compulsory license under this section. The regulations covering both the monthly and the annual statements of account shall prescribe the form, content, and manner of certification with respect to the number of records made and the number of records distributed.

(4) If the copyright owner does not receive the monthly payment and the monthly and annual statements of account when due, the owner may give written notice to the licensee that, unless the default is remedied within thirty days from the date of the notice, the compulsory license will be automatically terminated. Such termination renders either the making or the distribution, or both, of all phonorecords for which

the royalty has not been paid, actionable as acts of infringement under section 501 and fully subject to the remedies provided by sections 502 through 506 and 509.

§ 116. Scope of Exclusive Rights in Nondramatic Musical Works: Public Performances by Means of Coin-Operated Phonorecord Players

(a) *Limitation on Exclusive Right*

In the case of a nondramatic musical work embodied in a phonorecord, the exclusive right under clause (4) of section 106 to perform the work publicly by means of a coin-operated phonorecord player is limited as follows:

(1) The proprietor of the establishment in which the public performance takes place is not liable for infringement with respect to such public performance unless—

    (A) such proprietor is the operator of the phonorecord player; or

    (B) such proprietor refuses or fails, within one month after receipt by registered or certified mail of a request, at a time during which the certificate required by clause (1)(C) of subsection (b) is not affixed to the phonorecord player, by the copyright owner, to make full disclosure, by registered or certified mail, of the identity of the operator of the phonorecord player.

(2) The operator of the coin-operated phonorecord player may obtain a compulsory license to perform the work publicly on that phonorecord player by filing the application, affixing the certificate, and paying the royalties provided by subsection (b).

(b) *Recordation of Coin-Operated Phonorecord Player, Affixation of Certificate, and Royalty Payable under Compulsory License*

(1) Any operator who wishes to obtain a compulsory license for the public performance of works on a coin-operated phonorecord player shall fulfill the following requirements:

    (A) Before or within one month after such performances are made available on a particular phonorecord player, and during the month of January in each succeeding year that such performances are made available on that particular phonorecord player, the operator shall file in the Copyright Office, in accordance with requirements that the Register of Copyrights, after consultation with the Copyright Royalty Tribunal (if and when the Tribunal has been constituted), shall prescribe by regulation, an application containing the name and address of the operator of the phono-record player and the manufacturer and serial number or other explicit identification of the phonorecord player, and deposit with the Register of Copyrights a royalty fee for the current calendar year of $8 for that particular phonorecord player. If such performances are made available on a particular phonorecord player for the first time after July 1 of any year, the royalty fee to be deposited for the remainder of that year shall be $4.

(B) Within twenty days of receipt of an application and a royalty fee pursuant to subclause (A), the Register of Copyrights shall issue to the applicant a certificate for the phonorecord player.

(C) On or before March 1 of the year in which the certificate prescribed by subclause (B) of this clause is issued, or within ten days after the date of issue of the certificate, the operator shall affix to the particular phonorecord player, in a position where it can be readily examined by the public, the certificate, issued by the Register of Copyrights under subclause (B), of the latest application made by such operator under subclause (A) of this clause with respect to that phonorecord player.

(2) Failure to file the application, to affix the certificate, or to pay the royalty required by clause (1) of this subsection renders the public performance actionable as an act of infringement under section 501 and fully subject to the remedies provided by sections 502 through 506 and 509

(c) *Distribution of Royalties*

(1) The Register of Copyrights shall receive all fees deposited under this section and, after deducting the reasonable costs incurred by the Copyright Office under this section, shall deposit the balance in the Treasury of the United States, in such manner as the Secretary of the Treasury directs. All funds held by the Secretary of the Treasury shall be invested in interest-bearing United States securities for later distribution with interest by the Copyright Royalty Tribunal as provided by this title. The Register shall submit to the Copyright Royalty Tribunal, on an annual basis, a detailed statement of account covering all fees received for the relevant period provided by subsection (b).

(2) During the month of January in each year, every person claiming to be entitled to compulsory license fees under this section for performances during the preceding twelve-month period shall file a claim with the Copyright Royalty Tribunal, in accordance with requirements that the Tribunal shall prescribe by regulation. Such claim shall include an agreement to accept as final, except as provided in section 810 of this

title, the determination of the Copyright Royalty Tribunal in any controversy concerning the distribution of royalty fees deposited under subclause (A) of subsection (b)(1) of this section to which the claimant is a party. Notwithstanding any provisions of the antitrust laws, for purposes of this subsection any claimants may agree among themselves as to the proportionate division of compulsory licensing fees among them, may lump their claims together and file them jointly or as a single claim, or may designate a common agent to receive payment on their behalf.

(3) After the first day of October of each year, the Copyright Royalty Tribunal shall determine whether there exists a controversy concerning the distribution of royalty fees deposited under subclause (A) of subsection (b)(1). If the Tribunal determines that no such controversy exists, it shall, after deducting its reasonable administrative costs under this section, distribute such fees to the copyright owners entitled, or to their designated agents. If it finds that such a controversy exists, it shall, pursuant to chapter 8 of this title, conduct a proceeding to determine the distribution of royalty fees.

(4) The fees to be distributed shall be divided as follows:

(A) to every copyright owner not affiliated with a performing rights society, the pro rata share of the fees to be distributed to which such copyright owner proves entitlement.

(B) to the performing rights societies, the remainder of the fees to be distributed in such pro rata shares as they shall by agreement stipulate among themselves, or, if they fail to agree, the pro rata share to which such performing rights societies prove entitlement.

(C) during the pendency of any proceeding under this section, the Copyright Royalty Tribunal shall withhold from distribution an amount sufficient to satisfy all claims with respect to which a controversy exists, but shall have discretion to proceed to distribute any amounts that are not in controversy.

(5) The Copyright Royalty Tribunal shall promulgate regulations under which persons who can reasonably be expected to have claims may, during the year in which performances take place, without expense to or harassment of operators or proprietors of establishments in which phonorecord players are located, have such access to such establishments and to the phonorecord players located therein and such opportunity to obtain information with respect thereto as may be reasonably necessary to determine, by sampling procedures or otherwise, the proportion of contribution of the musical works of each such person to the earnings of the phonorecord players for which fees shall have been deposited. Any person who alleges that he or she has been denied the access permitted under the regulations prescribed by the Copyright Royalty Tribunal may bring an action in the United States District Court for

the District of Columbia for the cancellation of the compulsory license of the phonorecord player to which such access has been denied, and the court shall have the power to declare the compulsory license thereof invalid from the date of issue thereof.

(d) Criminal Penalties

Any person who knowingly makes a false representation of a material fact in an application filed under clause (1)(A) of subsection (b), or who knowingly alters a certificate issued under clause (1)(B) of subsection (b) or knowingly affixes such a certificate to a phonorecord player other than the one it covers, shall be fined not more than $2,500.

(e) Definitions

As used in this section, the following terms and their variant forms mean the following:

(1) A "coin-operated phonorecord player" is a machine or device that—

(A) is employed solely for the performance of nondramatic musical works by means of phonorecords upon being activated by insertion of coins, currency, tokens, or other monetary units or their equivalent;

(B) is located in an establishment making no direct or indirect charge for admission;

(C) is accompanied by a list of the titles of all the musical works available for performance on it, which list is affixed to the phonorecord player or posted in the establishment in a prominent position where it can be readily examined by the public; and

(D) affords a choice of works available for performance and permits the choice to be made by the patrons of the establishment in which it is located.

(2) An "operator" is any person who, alone or jointly with others:

(A) owns a coin-operated phonorecord player; or

(B) has the power to make a coin-operated phonorecord player available for placement in an establishment for purposes of public performance; or

(C) has the power to exercise primary control over the selection of the musical works made available for public performance on a coin-operated phonorecord player.

(3) A "performing rights society" is an association or corporation that licenses the public performance of nondramatic musical works on

behalf of the copyright owners, such as the American Society of Composers, Authors and Publishers, Broadcast Music, Inc., and SESAC, Inc.

§ 117. Scope of Exclusive Rights: Use in Conjunction with Computers and Similar Information Systems

Notwithstanding the provisions of sections 106 through 116 and 118, this title does not afford to the owner of copyright in a work any greater or lesser rights with respect to the use of the work in conjunction with automatic systems capable of storing, processing, retrieving, or transferring information, or in conjunction with any similar device, machine, or process, than those afforded to works under the law, whether title 17 or the common law or statutes of a State, in effect on December 31, 1977, as held applicable and construed by a court in an action brought under this title.

§ 118. Scope of Exclusive Rights: Use of Certain Works in Connection with Noncommercial Broadcasting

(a) The exclusive rights provided by section 106 shall, with respect to the works specified by subsection (b) and the activities specified by subsection (d), be subject to the conditions and limitations prescribed by this section.

(b) Not later than thirty days after the Copyright Royalty Tribunal has been constituted in accordance with section 802, the Chairman of the Tribunal shall cause notice to be published in the Federal Register of the initiation of proceedings for the purpose of determining reasonable terms and rates of royalty payments for the activities specified by subsection (d) with respect to published nondramatic musical works and published pictorial, graphic, and sculptural works during a period beginning as provided in clause (3) of this subsection and ending on December 31, 1982. Copyright owners and public broadcasting entities shall negotiate in good faith and cooperate fully with the Tribunal in an effort to reach reasonable and expeditious results. Notwithstanding any provision of the antitrust laws, any owners of copyright in works specified by this subsection and any public broadcasting entities, respectively, may negotiate and agree upon the terms and rates of royalty payments and the proportionate division of fees paid among various copyright owners, and may designate common agents to negotiate, agree to, pay, or receive payments.

(1) Any owner of copyright in a work specified in this subsection or any public broadcasting entity may, within one hundred and twenty days after publication of the notice specified in this subsection, submit to the

Copyright Royalty Tribunal proposed licenses covering such activities with respect to such works. The Copyright Royalty Tribunal shall proceed on the basis of the proposals submitted to it as well as any other relevant information. The Copyright Royalty Tribunal shall permit any interested party to submit information relevant to such proceedings.

(2) License agreements voluntarily negotiated at any time between one or more copyright owners and one or more public broadcasting entities shall be given effect in lieu of any determination by the Tribunal: *Provided,* That copies of such agreements are filed in the Copyright Office within thirty days of execution in accordance with regulations that the Register of Copyrights shall prescribe.

(3) Within six months, but not earlier than one hundred and twenty days, from the date of publication of the notice specified in this subsection the Copyright Royalty Tribunal shall make a determination and publish in the Federal Register a schedule of rates and terms which, subject to clause (2) of this subsection, shall be binding on all owners of copyright in works specified by this subsection and public broadcasting entities, regardless of whether or not such copyright owners and public broadcasting entities have submitted proposals to the Tribunal. In establishing such rates and terms the Copyright Royalty Tribunal may consider the rates for comparable circumstances under voluntary license agreements negotiated as provided in clause (2) of this subsection. The Copyright Royalty Tribunal shall also establish requirements by which copyright owners may receive reasonable notice of the use of their works under this section, and under which records of such use shall be kept by public broadcasting entities.

(4) With respect to the period beginning on the effective date of this title and ending on the date of publication of such rates and terms, this title shall not afford to owners of copyright or public broadcasting entities any greater or lesser rights with respect to the activities specified in subsection (d) as applied to works specified in this subsection than those afforded under the law in effect on December 31, 1977, as held applicable and construed by a court in an action brought under this title.

(c) The initial procedure specified in subsection (b) shall be repeated and concluded between June 30 and December 31, 1982, and at five-year intervals thereafter, in accordance with regulations that the Copyright Royalty Tribunal shall prescribe.

(d) Subject to the transitional provisions of subsection (b)(4), and to the terms of any voluntary license agreements that have been negotiated as provided by subsection (b)(2), a public broadcasting entity may, upon compliance with the provisions of this section, including the rates and terms established by the Copyright Royalty Tribunal under subsection (b)(3), engage in the following activities with respect to published nondramatic musical works and published pictorial, graphic, and sculptural works:

(1) performance or display of a work by or in the course of a transmission made by a noncommercial educational broadcast station referred to in subsection (g); and

(2) production of a transmission program, reproduction of copies or phonorecords of such a transmission program, and distribution of such copies or phonorecords, where such production, reproduction, or distribution is made by a nonprofit institution or organization solely for the purpose of transmissions specified in clause (1); and

(3) the making of reproductions by a governmental body or a nonprofit institution of a transmission program simultaneously with its transmission as specified in clause (1), and the performance or display of the contents of such program under the conditions specified by clause (1) of section 110, but only if the reproductions are used for performances or displays for a period of no more than seven days from the date of the transmission specified in clause (1), and are destroyed before or at the end of such period. No person supplying, in accordance with clause (2), a reproduction of a transmission program to governmental bodies or nonprofit institutions under this clause shall have any liability as a result of failure of such body or institution to destroy such reproduction: *Provided,* That it shall have notified such body or institution of the requirement for such destruction pursuant to this clause: *And provided further,* That if such body or institution itself fails to destroy such reproduction it shall be deemed to have infringed.

(e) Except as expressly provided in this subsection, this section shall have no applicability to works other than those specified in subsection (b).

(1) Owners of copyright in nondramatic literary works and public broadcasting entities may, during the course of voluntary negotiations, agree among themselves, respectively, as to the terms and rates of royalty payments without liability under the antitrust laws. Any such terms and rates of royalty payments shall be effective upon filing in the Copyright Office, in accordance with regulations that the Register of Copyrights shall prescribe.

(2) On January 3, 1980, the Register of Copyrights, after consulting with authors and other owners of copyright in nondramatic literary works and their representatives, and with public broadcasting entities and their representatives, shall submit to the Congress a report setting forth the extent to which voluntary licensing arrangements have been reached with respect to the use of nondramatic literary works by such broadcast stations. The report should also describe any problems that

may have arisen, and present legislative or other recommendations, if warranted.

(f) Nothing in this section shall be construed to permit, beyond the limits of fair use as provided by section 107, the unauthorized dramatization of a nondramatic musical work, the production of a transmission program drawn to any substantial extent from a published compilation of pictorial, graphic, or sculptural works, or the unauthorized use of any portion of an audiovisual work.

(g) As used in this section, the term "public broadcasting entity" means a noncommercial educational broadcast station as defined in section 397 of title 47 and any nonprofit institution or organization engaged in the activities described in clause (2) of subsection (d).

§ 301. Preemption with Respect to Other Laws

(a) On and after January 1, 1978, all legal or equitable rights that are equivalent to any of the exclusive rights within the general scope of copyright as specified by section 106 in works of authorship that are fixed in a tangible medium of expression and come within the subject matter of copyright as specified by sections 102 and 103, whether created before or after that date and whether published or unpublished, are governed exclusively by this title. Thereafter, no person is entitled to any such right or equivalent right in any such work under the common law or statutes of any State.

(b) Nothing in this title annuls or limits any rights or remedies under the common law or statutes of any State with respect to—

(1) subject matter that does not come within the subject matter of copyright as specified by sections 102 and 103, including works of authorship not fixed in any tangible medium of expression; or

(2) any cause of action arising from undertakings commenced before January 1, 1978; or

(3) activities violating legal or equitable rights that are not equivalent to any of the exclusive rights within the general scope of copyright as specified by section 106.

·   ·   ·   ·   ·

§ 504. Remedies for Infringement: Damages and Profits

(c) *Statutory Damages*

·   ·   ·   ·   ·

(2) In a case where the copyright owner sustains the burden of

proving, and the court finds, that infringement was committed willfully, the court in its discretion may increase the award of statutory damages to a sum of not more than $50,000. In a case where the infringer sustains the burden of proving, and the court finds, that such infringer was not aware and had no reason to believe that his or her acts constituted an infringement of copyright, the court at its discretion may reduce the award of statutory damages to a sum of not less than $100. The court shall remit statutory damages in any case where an infringer believed and had reasonable grounds for believing that his or her use of the copyrighted work was a fair use under section 107, if the infringer was: (i) an employee or agent of a nonprofit educational institution, library, or archives acting within the scope of his or her employment who, or such institution, library, or archives itself, which infringed by reproducing the work in copies or phonorecords; or (ii) a public broadcasting entity which or a person who, as a regular part of the nonprofit activities of a public broadcasting entity (as defined in subsection (g) of section 118) infringed by performing a published nondramatic literary work or by reproducing a transmission program embodying a performance of such a work.

§ 602. Infringing Importation of Copies or Phonorecords

(a) Importation into the United States, without the authority of the owner of copyright under this title, of copies or phonorecords of a work that have been acquired outside the United States is an infringement of the exclusive right to distribute copies or phonorecords under section 106, actionable under section 501. This subsection does not apply to—

. . . . .

(3) importation by or for an organization operated for scholarly, educational, or religious purposes and not for private gain, with respect to no more than one copy of an audiovisual work solely for its archival purposes, and no more than five copies or phonorecords of any other work for its library lending or archival purposes, unless the importation of such copies or phonorecords is part of an activity consisting of systematic reproduction or distribution, engaged in by such organization in violation of the provisions of section 108(g)(2).

. . . . .

Appendix B

Committee Reports on Section 107, Fair Use

(1) House Committee Report on the 1976 Copyright Bill (House Committee on the Judiciary, House Report No. 94-1476 to accompany S. 22, 94th Cong., 2d Sess., September 3, 1976, pp. 65–74)

General Background of the Problem

The judicial doctrine of fair use, one of the most important and well-established limitations on the exclusive right of copyright owners, would be given express statutory recognition for the first time in section 107. The claim that a defendant's acts constituted a fair use rather than an infringement has been raised as a defense in innumerable copyright actions over the years, and there is ample case law recognizing the existence of the doctrine and applying it. The examples enumerated at page 24 of the Register's 1961 Report, while by no means exhaustive, give some idea of the sort of activities the courts might regard as fair use under the circumstances: "quotation of excerpts in a review or criticism for purposes of illustration or comment; quotation of short passages in a scholarly or technical work, for illustration or clarification of the author's observations; use in a parody of some of the content of the work parodied; summary of an address or article, with brief quotations, in a news report; reproduction by a library of a portion of a work to replace part of a damaged copy; reproduction by a teacher or student of a small part of a work to illustrate a lesson; reproduction of a work in legislative or judicial proceedings or reports; incidental and fortuitous reproduction, in a newsreel or broadcast, of a work located in the scene of an event being reported."

Although the courts have considered and ruled upon the fair use doctrine over and over again, no real definition of the concept has ever emerged. Indeed, since the doctrine is an equitable rule of reason, no generally applicable definition is possible, and each case raising the question must be decided on its own facts. On the other hand, the courts have evolved a set of criteria which, though in no case definitive or determinative, provide some gauge for balancing the equities. These criteria have been stated in various ways, but essentially they can all be reduced to the four standards which have been adopted in section 107: "(1) the purpose and character of the use, including whether such use is of a commercial nature or is for non-profit educational purposes; (2) the nature of the copyrighted work; (3) the amount and substantiality of the portion used in relation to the copyrighted work as a whole; and (4) the effect of the use upon the potential market for or value of the copyrighted work."

These criteria are relevant in determining whether the basic doctrine of fair use, as stated in the first sentence of section 107, applies in a particular case: "Notwithstanding the provisions of section 106, the fair use of a copyrighted work, including such use by reproduction in copies or phonorecords or by any other means specified by that section, for purposes such as criticism, comment, news reporting, teaching (including multiple copies for classroom use), scholarship, or research, is not an infringement of copyright."

The specific wording of section 107 as it now stands is the result of a process of accretion, resulting from the long controversy over the related problems of fair use and the reproduction (mostly by photocopying) of copyrighted material for educational and scholarly purposes. For example, the reference to fair use "by reproduction in copies or phonorecords or by any other means" is mainly intended to make clear that the doctrine has as much application to photocopying and taping as to older forms of use; it is not intended to give these kinds of reproduction any special status under the fair use provision or to sanction any reproduction beyond the normal and reasonable limits of fair use. Similarly, the newly-added reference to "multiple copies for classroom use" is a recognition that, under the proper circumstances of fairness, the doctrine can be applied to reproductions of multiple copies for the members of a class.

The Committee has amended the first of the criteria to be considered—"the purpose and character of the use"—to state explicitly that this factor includes a consideration of "whether such use is of a commercial nature or is for non-profit educational purposes." This amendment is not intended to be interpreted as any sort of not-for-profit limitation on educational uses of copyrighted works. It is an express recognition that, as under the present law, the commercial or non-profit character of an activity, while not conclusive with respect to fair use, can and should be weighed along with other factors in fair use decisions.

General Intention behind the Provision

The statement of the fair use doctrine in section 107 offers some guidance to users in determining when the principles of the doctrine apply. However, the endless variety of situations and combinations of circumstances that can rise in particular cases precludes the formulation of exact rules in the statute. The bill endorses the purpose and general scope of the judicial doctrine of fair use, but there is no disposition to freeze the doctrine in the statute, especially during a period of rapid technological change. Beyond a very broad statutory explanation of what fair use is and some of the criteria applicable to it, the courts must be free to adapt the doctrine to particular situations on a case-by-case basis. Section 107 is intended to restate the present judicial doctrine of fair use, not to change, narrow, or enlarge it in any way.

Intention as to Classroom Reproduction

Although the works and uses to which the doctrine of fair use is applicable are as broad as the copyright law itself, most of the discussion of section 107 has centered around questions of classroom reproduction, particularly photocopying. The arguments on the question are summarized at pp. 30–31 of this Committee's 1967 report (H.R. Rep. No. 83, 90th Cong., 1st Sess.), and have not changed materially in the intervening years.

The Committee also adheres to its earlier conclusion, that "a specific exemption freeing certain reproductions of copyrighted works for educational and scholarly purposes from copyright control is not justified." At the same time the Committee recognizes, as it did in 1967, that there is a "need for greater certainty and protection for teachers." In an effort to meet this need the Committee has not only adopted further amendments to section 107, but has also amended section 504(c) to provide innocent teachers and other non-profit users of copyrighted material with broad insulation against unwarranted liability for infringement. The latter amendments are discussed below in connection with Chapter 5 of the bill.

In 1967 the Committee also sought to approach this problem by including, in its report, a very thorough discussion of "the considerations lying behind the four criteria listed in the amended section 107, in the context of typical classroom situations arising today." This discussion appeared on pp. 32–35 of the 1967 report, and with some changes has been retained in the Senate report on S. 22 (S. Rep. No. 94–473, pp. 63–65). The Committee has reviewed this discussion, and considers that it still has value as an analysis of various aspects of the problem.

At the Judiciary Subcommittee hearings in June 1975, Chairman Kastenmeier and other members urged the parties to meet together independently in an effort to achieve a meeting of the minds as to permissible educational uses of copyrighted material. The response to these suggestions was positive, and a number of meetings of three groups, dealing respectively with classroom reproduction of printed material, music, and audio-visual material, were held beginning in September 1975.

In a joint letter to Chairman Kastenmeier, dated March 19, 1976, the representatives of the Ad Hoc Committee of Educational Institutions and Organizations on Copyright Law Revision, and of the Authors League of America, Inc., and the Association of American Publishers, Inc., stated:

> You may remember that in our letter of March 8, 1976 we told you that the negotiating teams representing authors and publishers and the Ad Hoc Group had reached tentative agreement on guidelines to insert in the Committee Report covering educational copying from books and periodicals under Section 107 of H.R. 2223 and S. 22, and that as part of that tentative agreement each side would accept the amendments to Sections 107 and 504 which were adopted by your Subcommittee on March 3, 1976.
>
> We are now happy to tell you that the agreement has been approved by the principals and we enclose a copy herewith. We had originally intended to translate the agreement into language suitable for inclusion in the legislative report dealing with Section 107, but we have since been advised by committee staff that this will not be necessary.
>
> As stated above, the agreement refers only to copying from books and periodicals, and it is not intended to apply to musical or audiovisual works.

The full text of the agreement is as follows:

AGREEMENT ON GUIDELINES FOR CLASSROOM COPYING IN NOT-FOR-PROFIT EDUCATIONAL INSTITUTIONS

WITH RESPECT TO BOOKS AND PERIODICALS

The purpose of the following guidelines is to state the minimum and not the maximum standards of educational fair use under Section 107 of H.R. 2223. The parties agree that the conditions determining the extent of permissible copying for educational purposes may change in the future; that certain types of copying permitted under these guidelines may not be permissible in the future; and conversely that in the future other types of copying not permitted under these guidelines may be permissible under revised guidelines.

Moreover, the following statement of guidelines is not intended to limit the types of copying permitted under the standards of fair use under judicial decision and which are stated in Section 107 of the Copyright Revision Bill. There may be instances in which copying which does not fall within the guidelines stated below may nonetheless be permitted under the criteria of fair use.

GUIDELINES

I. *Single Copying for Teachers*

A single copy may be made of any of the following by or for a teacher at his or her individual request for his or her scholarly research or use in teaching or preparation to teach a class:

A. A chapter from a book;

B. An article from a periodical or newspaper;

C. A short story, short essay or short poem, whether or not from a collective work;

D. A chart, graph, diagram, drawing, cartoon or picture from a book, periodical, or newspaper;

II. *Multiple Copies for Classroom Use*

Multiple copies (not to exceed in any event more than one copy per pupil in a course) may be made by or for the teacher giving the course for classroom use or discussion; *provided that:*

A. The copying meets the tests of brevity and spontaneity as defined below; *and,*

B. Meets the cumulative effect test as defined below; *and,*

C. Each copy includes a notice of copyright

Definitions

Brevity

(*i*) Poetry: (a) A complete poem if less than 250 words and if printed on not more than two pages or, (b) from a longer poem, an excerpt of not more than 250 words.

(*ii*) Prose: (a) Either a complete article, story or essay of less than 2,500 words, or (b) an excerpt from any prose work of not more than 1,000 words or 10% of the work, whichever is less, but in any event a minimum of 500 words.

[Each of the numerical limits stated in "i" and "ii" above may be expanded to permit the completion of an unfinished line of a poem or of an unfinished prose paragraph.]

(*iii*) Illustration: One chart, graph, diagram, drawing, cartoon or picture per book or per periodical issue.

(*iv*) "Special" works: Certain works in poetry, prose or in "poetic prose" which often combine language with illustrations and which are intended sometimes for children and at other times for a more general audience fall short of 2,500 words in their entirety. Paragraph "ii" above notwithstanding such "special works" may not be reproduced in their entirety; however, an excerpt comprising not more than two of the published pages of such special work and containing not more than 10% of the words found in the text thereof, may be reproduced.

Spontaneity

(*i*) The copying is at the instance and inspiration of the individual teacher, and

(*ii*) The inspiration and decision to use the work and the moment of its use for maximum teaching effectiveness are so close in time that it would be unreasonable to expect a timely reply to a request for permission.

Cumulative Effect

(*i*) The copying of the material is for only one course in the school in which the copies are made.

(*ii*) Not more than one short poem, article, story, essay or two excerpts may be copied from the same author, nor more than three from the same collective work or periodical volume during one class term.

(*iii*) There shall not be more than nine instances of such multiple copying for one course during one class term.

[The limitations stated in "ii" and "iii" above shall not apply to current news periodicals and newspapers and current news sections of other periodicals.]

III. *Prohibitions as to I and II Above*

Notwithstanding any of the above, the following shall be prohibited:

(A) Copying shall not be used to create or to replace or substitute for anthologies, compilations or collective works. Such replacement or substitution may occur whether copies of various works or excerpts therefrom are accumulated or reproduced and used separately.

(B) There shall be no copying of or from works intended to be "consumable" in the course of study or of teaching. These include workbooks, exercises, standardized tests and test booklets and answer sheets and like consumable material.

(C) Copying shall not:

(a) substitute for the purchase of books, publishers' reprints or periodicals;

(b) be directed by higher authority;

(c) be repeated with respect to the same item by the same teacher from term to term.

(D) No charge shall be made to the student beyond the actual cost of the photocopying.

Agreed MARCH 19, 1976.

Ad Hoc Committee on Copyright Law Revision:

By SHELDON ELLIOTT STEINBACH.

Author-Publisher Group:

Authors League of America:

By IRWIN KARP, *Counsel.*

Association of American Publishers, Inc.:

By ALEXANDER C. HOFFMAN,
Chairman, Copyright Committee.

In a joint letter dated April 30, 1976, representatives of the Music Publishers' Association of the United States, Inc., the National Music

Publishers' Association, Inc., the Music Teachers National Association, the Music Educators National Conference, the National Association of Schools of Music, and the Ad Hoc Committee on Copyright Law Revision, wrote to Chairman Kastenmeier as follows:

> During the hearings on H.R. 2223 in June 1975, you and several of your subcommittee members suggested that concerned groups should work together in developing guidelines which would be helpful to clarify Section 107 of the bill.
>
> Representatives of music educators and music publishers delayed their meetings until guidelines had been developed relative to books and periodicals. Shortly after that work was completed and those guidelines were forwarded to your subcommittee, representatives of the undersigned music organizations met together with representatives of the Ad Hoc Committee on Copyright Law Revision to draft guidelines relative to music.
>
> We are very pleased to inform you that the discussions thus have been fruitful on the guidelines which have been developed. Since private music teachers are an important factor in music education, due consideration has been given to the concerns of that group.
>
> We trust that this will be helpful in the report on the bill to clarify Fair Use as it applies to music.

The text of the guidelines accompanying this letter is as follows:

GUIDELINES FOR EDUCATIONAL USES OF MUSIC

The purpose of the following guidelines is to state the minimum and not the maximum standards of educational fair use under Section 107 of HR 2223. The parties agree that the conditions determining the extent of permissible copying for educational purposes may change in the future; that certain types of copying permitted under these guidelines may not be permissible in the future, and conversely that in the future other types of copying not permitted under these guidelines may be permissible under revised guidelines.

Moreover, the following statement of guidelines is not intended to limit the types of copying permitted under the standards of fair use under judicial decision and which are stated in Section 107 of the Copyright Revision Bill. There may be instances in which copying which does not fall within the guidelines stated below may nonetheless be permitted under the criteria of fair use.

A. Permissible Uses

1. Emergency copying to replace purchased copies which for any reason are not available for an imminent performance provided purchased replacement copies shall be substituted in due course.

2. (a) For academic purposes other than performance, single or multiple copies of excerpts of works may be made, provided that the excerpts do not comprise a part of the whole which would constitute a performable unit such as a section, movement or aria, but in no case more than 10% of the whole work. The number of copies shall not exceed one copy per pupil.

(b) For academic purposes other than performance, a single copy of an entire performable unit (section, movement, aria, etc.) that is, (1) confirmed by the copyright proprietor to be out of print or (2) unavailable except in a larger work, may be made by or for a teacher solely for the purpose of his or her scholarly research or in preparation to teach a class.

3. Printed copies which have been purchased may be edited or simplified provided that the fundamental character of the work is not distorted or the lyrics, if any, altered or lyrics added if none exist.

4. A single copy of recordings of performances by students may be made for evaluation or rehearsal purposes and may be retained by the educational institution or individual teacher.

5. A single copy of a sound recording (such as a tape, disc or cassette) of copyrighted music may be made from sound recordings owned by an educational institution or an individual teacher for the purpose of constructing aural exercises or examinations and may be retained by the educational institution or individual teacher. (This pertains only to the copyright of the music itself and not to any copyright which may exist in the sound recording.)

B. Prohibitions

1. Copying to create or replace or substitute for anthologies, compilations or collective works.

2. Copying of or from works intended to be "consumable" in the course of study or of teaching such as workbooks, exercises, standardized tests and answer sheets and like material.

3. Copying for the purpose of performance, except as in A(1) above.

4. Copying for the purpose of substituting for the purchase of music, except as in A(1) and A(2) above.

5. Copying without inclusion of the copyright notice which appears on the printed copy.

The problem of off-the-air taping for nonprofit classroom use of copyrighted audiovisual works incorporated in radio and television broadcasts has proved to be difficult to resolve. The Committee believes that the fair use doctrine has some limited application in this area, but it appears that the development of detailed guidelines will require a more thorough exploration than has so far been possible of the needs and problems of a number of different interests affected, and of the various legal problems presented. Nothing in section 107 or elsewhere in the bill is intended to change or prejudge the law on the point. On the other hand, the Committee is sensitive to the importance of the problem, and urges the representatives of the various interests, if possible under the leadership of the Register of Copyrights, to continue their discussions actively and in a constructive spirit. If it would be helpful to a solution, the Committee is receptive to undertaking further consideration of the problem in a future Congress.

The Committee appreciates and commends the efforts and the cooperative and reasonable spirit of the parties who achieved the agreed guidelines on books and periodicals and on music. Representatives of the American Association of University Professors and of the Association of American Law Schools have written to the Committee strongly criticizing the guidelines, particularly with respect to multiple copying, as being too restrictive with respect to classroom situations at the university and graduate level. However, the Committee notes that the Ad Hoc group did include representatives of higher education, that the stated "purpose of the . . . guidelines is to state the minimum and not the maximum standards of educational fair use" and that the agreement acknowledges "there may be instances in which copying which does not fall within the guidelines . . . may nonetheless be permitted under the criteria of fair use."

The Committee believes the guidelines are a reasonable interpretation of the minimum standards of fair use. Teachers will know that copying within the guidelines is fair use. Thus, the guidelines serve the purpose of fulfilling the need for greater certainty and protection for teachers. The Committee expresses the hope that if there are areas where standards other than these guidelines may be appropriate, the parties will continue their efforts to provide additional specific guidelines in the same spirit of good will and give and take that has marked the discussion of this subject in recent months.

Reproduction and Uses for Other Purposes

The concentrated attention given the fair use provision in the context of classroom teaching activities should not obscure its application in other areas. It must be emphasized again that the same general standards of fair use are applicable to all kinds of uses of copyrighted material, although the relative weight to be given them will differ from case to case.

The fair use doctrine would be relevant to the use of excerpts from copyrighted works in educational broadcasting activities not exempted under section 110(2) or 112, and not covered by the licensing provisions of section 118. In these cases the factors to be weighed in applying the criteria of this section would include whether the performers, producers, directors, and others responsible for the broadcast were paid, the size and nature of the audience, the size and number of excerpts taken and, in the case of recordings made for broadcast, the number of copies reproduced and the extent of their reuse or exchange. The availability of the fair use doctrine to educational broadcasters would be narrowly circumscribed in the case of motion pictures and other audiovisual works, but under appropriate circumstances it could apply to the nonsequential showing of an individual still or slide, or to the performance of a short excerpt from a motion picture for criticism or comment.

Another special instance illustrating the application of the fair use doctrine pertains to the making of copies or phonorecords of works in the special forms needed for the use of blind persons. These special forms, such as copies in Braille and phonorecords of oral readings (talking books), are not usually made by the publishers for commercial distribution. For the most part, such copies and phonorecords are made by the Library of Congress' Division for the Blind and Physically Handicapped with permission obtained from the copyright owners, and are circulated to blind persons through regional libraries covering the nation. In addition, such copies and phonorecords are made locally by individual volunteers for the use of blind persons in their communities, and the Library of Congress conducts a program for training such volunteers. While the making of multiple copies or phonorecords of a work for general circulation requires the permission of the copyright owner, a problem addressed in section 710 of the bill, the making of a single copy or phonorecord by an individual as a free service for a blind persons would properly be considered a fair use under section 107.

A problem of particular urgency is that of preserving for posterity prints of motion pictures made before 1942. Aside from the deplorable fact that in a great many cases the only existing copy of a film has been deliberately destroyed, those that remain are in immediate danger of disintegration; they were printed on film stock with a nitrate base that will inevitably decompose in time. The efforts of the Library of Congress, the American Film Institute, and other organizations to rescue and preserve this irreplaceable contribution to our cultural life are to be applauded, and the making of duplicate copies for purposes of archival preservation certainly falls within the scope of "fair use."

When a copyrighted work contains unfair, inaccurate, or derogatory information concerning an individual or institution, the individual or institution may copy and reproduce such parts of the work as are necessary to permit understandable comment on the statements made in the work.

The Committee has considered the question of publication, in Congressional hearings and documents, of copyrighted material. Where the length of the work or excerpt published and the number of copies authorized are reasonable under the circumstances, and the work itself is directly relevant to a matter of legitimate legislative concern, the Committee believes that the publication would constitute fair use.

During the consideration of the revision bill in the 94th Congress it was proposed that independent newsletters, as distinguished from house organs and publicity or advertising publications, be given separate treatment. It is argued that newsletters are particularly vulnerable to mass photocopying, and that most newsletters have fairly modest circulations.

Whether the copying of portions of a newsletter is an act of infringement or a fair use will necessarily turn on the facts of the individual case. However, as a general principle, it seems clear that the scope of the fair use doctrine should be considerably narrower in the case of newsletters than in that of either mass-circulation periodicals or scientific journals. The commercial nature of the user is a significant factor in such cases: Copying by a profit-making user of even a small portion of a newsletter may have a significant impact on the commercial market for the work.

The Committee has examined the use of excerpts from copyrighted works in the art work of calligraphers. The committee believes that a single copy reproduction of an excerpt from a copyrighted work by a calligrapher for a single client does not represent an infringement of copyright. Likewise, a single reproduction of excerpts from a copyrighted work by a student calligrapher or teacher in a learning situation would be a fair use of the copyrighted work.

The Register of Copyrights has recommended that the committee report describe the relationship between this section and the provisions of section 108 relating to reproduction by libraries and archives. The doctrine of fair use applies to library photocopying, and nothing contained in section 108 "in any way affects the right of fair use." No provision of section 108 is intended to take away any rights existing under the fair use doctrine. To the contrary, section 108 authorizes certain photocopying practices which may not qualify as a fair use.

The criteria of fair use are necessarily set forth in general terms. In the application of the criteria of fair use to specific photocopying practices of libraries, it is the intent of this legislation to provide an appropriate balancing of the rights of creators, and the needs of users.

(2) Senate Committee Report on the 1976 Copyright Bill (Senate Committee on the Judiciary, Senate Report No. 94-473 to accompany S. 22, 94th Cong., 1st Sess., November 20, 1975, pp. 61–67)

[The first two paragraphs are identical to those of the House Committee Report.]

·　　·　　·　　·　　·

The underlying intention of the bill with respect to the application of the fair use doctrine in various situations is discussed below. It should be emphasized again that, in those situations or any others, there is no purpose of either freezing or changing the doctrine. In particular, the reference to fair use "by reproduction in copies or phonorecords or by any other means" should not be interpreted as sanctioning any reproduction beyond the normal and reasonable limits of fair use. In making separate mention of "reproduction in copies or phonorecords" in the section, the provision is not intended to give this kind of use any special or preferred status as compared with other kinds of uses. In any event, whether a use referred to in the first sentence of section 107 is a fair use in a particular case will depend upon the application of the determinative factors, including those mentioned in the second sentence.

Intention behind the Provision

In general. The statement of the fair use doctrine in section 107 offers some guidance to users in determining when the principles of the doctrine apply. However, the endless variety of situations and combinations of circumstances that can rise in particular cases precludes the formulation of exact rules in the statute. The bill endorses the purpose and general scope of the judicial doctrine of fair use, as outlined earlier in this report, but there is no disposition to freeze the doctrine in the statute, especially during a period of rapid technological change. Beyond a very broad statutory explanation of what fair use is and some of the criteria applicable to it, the courts must be free to adapt the doctrine to particular situations on a case-by-case basis.

Section 107 is intended to restate the present judicial doctrine of fair use, not to change, narrow, or enlarge it in any way. However, since this section will represent the first statutory recognition of the doctrine in our copyright law, some explanation of the considerations behind the language used in the list of four criteria is advisable. This is particularly true as to cases of copying by teachers, and by public libraries, since in these areas there are few if any judicial guidelines.

The statements in this report with respect to each of the criteria of fair use are necessarily subject to qualifications, because they must be applied in combination with the circumstances pertaining to other criteria, and because new conditions arising in the future may alter the balance of equities. It is also important to emphasize that the singling out of some instances to discuss in the context of fair use is not intended to indicate that other activities would or would not be beyond fair use.

The Purpose and Nature of the Use

Copyright recognized. Section 107 makes it clear that, assuming the applicable criteria are met, fair use can extend to the reproduction of copyrighted material for purposes of classroom teaching.

Nonprofit element. Although it is possible to imagine situations in which use by a teacher in an educational organization operated for profit (day camps, language schools, business schools, dance studios, et cetera) would constitute a fair use, the nonprofit character of the school in which the teacher works should be one factor to consider in determining fair use. Another factor would be whether any charge is made for the copies distributed.

Spontaneity. The fair use doctrine in the case of classroom copying would apply primarily to the situation of a teacher who, acting individually and at his own volition, makes one or more copies for temporary use by himself or his pupils in the classroom. A different result is indicated where the copying was done by the educational institution, school system, or larger unit or where copying was required or suggested by the school administration, either in special instances or as part of a general plan.

Single and multiple copying. Depending upon the nature of the work and other criteria, the fair use doctrine should differentiate between the amount of work that can be reproduced by a teacher for his own classroom use (for example, for reading or projecting a copy or for playing a tape recording), and the amount that can be reproduced for distribution to pupils. In the case of multiple copies, other factors would be whether the number reproduced was limited to the size of the class, whether circulation beyond the classroom was permitted, and whether the copies were recalled or destroyed after temporary use.

Collection and anthologies. Spontaneous copying of an isolated extract by a teacher, which may be a fair use under appropriate circumstances, could turn into an infringement if the copies were accumulated over a period of time with other parts of the same work, or were collected with other material from various works so as to constitute an anthology.

Special uses. There are certain classroom uses which, because of their special nature, would not be considered an infringement in the ordinary case. For example, copying of extracts by pupils as exercises in a shorthand or typing class or for foreign language study, or recordings of performances by music students for purposes of analysis and criticism, would normally be regarded as a fair use unless the copies of phonorecords were retained or duplicated.

The Nature of the Copyrighted Work

Character of the work. The character and purpose of the work will have a lot to do with whether its reproduction for classroom purposes is fair use or infringement. For example, in determining whether a teacher could make one or more copies without permission, a news article from the daily press would be judged differently from a full orchestral score of a musical composition. In general terms it could be expected that the doctrine of fair use would be applied strictly to the classroom reproduction of entire works, such as musical compositions, dramas, and audiovisual works including motion pictures, which by their nature are intended for performance or public exhibition.

Similarly, where the copyright work is intended to be "consumable" in the course of classroom activities—workbooks, exercise, standardized tests, and answer sheets are examples—the privilege of fair use by teachers or pupils would have little if any application. Text books and other material prepared primarily for the school markets would be less susceptible to reproduction for classroom use than material prepared for general public distribution. With respect to material in newspapers and' periodicals the doctrine of fair use should be liberally applied to allow copying of items of current interest to supplement and update the students' textbooks, but this would not extend to copying from periodicals published primarily for student use.

Availability of the work. A key, though not necessarily determinative, factor in fair use is whether or not the work is available to the potential user. If the work is "out of print" and unavailable for purchase through normal channels, the user may have more justification for reproducing it than in the ordinary case, but the existence of organizations licensed to provide photocopies of out-of-print works at reasonable cost is a factor to be considered. The applicability of the fair use doctrine to unpublished works is narrowly limited since, although the work is unavailable, this is the result of a deliberate choice on the part of the copyright owner. Under ordinary circumstances the copyright owner's "right of first publication" would outweigh any needs of reproduction for classroom purposes.

The Amount and Substantiality of the Material Used

During the consideration of this legislation there has been considerable discussion of the difference between an "entire work" and an "excerpt". The educators have sought a limited right for a teacher to make a single copy of an "entire" work for classroom purposes, but it seems apparent that this was not generally intended to extend beyond a "separately cognizable" or "self-contained" portion (for example a single poem, story, or article) in a collective work, and that no privilege is sought to reproduce an entire collective work (for example, an encyclopedia volume, a periodical issue) or a sizable integrated work published as an entity (a novel, treatise, monograph, and so forth). With this limitation, and subject to the other revelant criteria, the requested privilege of making a single copy appears appropriately to be within the scope of fair use.

The educators also sought statutory authority for the privilege of making "a reasonable number of copies or phonorecords for excerpts or quotations * * *, provided such excerpts or quotations are not substantial in length in proportion to their source." In general, and assuming the other necessary factors are present, the copying for classroom purposes of extracts or portions, which are not self-contained and which are relatively "not substantial in length" when compared to the larger, self-contained work from which they are taken, should be considered fair use. Depending on the circumstances, the same would also be true of very short self-contained works such as a brief poem, a map in a newspaper, a "vocabulary builder" from a monthly magazine, and so forth. This should not be construed as permitting a teacher to make multiple copies of the same work on a repetitive basis or for continued use.

Effect of Use on Potential Market for or Value of Work

This factor must almost always be judged in conjunction with the other three criteria. With certain special exceptions (use in parodies or as evidence in court proceedings might be examples) a use that supplants any part of the normal market for a copyrighted work would ordinarily be considered an infringement. As in any other case, whether this would be the result of reproduction by a teacher for classroom purposes requires an evaluation of the nature and purpose of the use, the type of work involved, and the size and relative importance of the portion taken. Fair use is essentially supplementary by nature, and classroom copying that exceeds the legitimate teaching aims such as filling in missing information or bringing a subject up to date would go beyond the proper bounds of fair use. Isolated instances of minor infringements, when multiplied many times, become in the aggregate a major inroad on copyright that must be prevented.

Reproduction and Uses for Other Purposes

The concentrated attention given the fair use provision in the context of classroom teaching activities should not obscure its application in other areas. It must be emphasized again that the same general standards of fair use are applicable to all kinds of uses of copyrighted material, although the relative weight to be given them will differ from case to case.

The fair use doctrine would be relevant to the use of excerpts from copyrighted works in educational broadcasting activities not exempted under section 110(2) or 112. In these cases the factors to be weighed in applying the criteria of this section would include whether the performers, producers, directors, and others responsible for the broadcast were paid, the size and nature of the audience, the size and number of excerpts taken and, in the case of recordings made for broadcast, the number of copies reproduced and the extent of their reuse or exchange. The availability of the fair use doctrine to educational broadcasters would be narrowly circumscribed in the case of motion pictures and other audiovisual works, but under appropriate circumstances it could apply to the nonsequential showing of an individual still or slide, or to the performance of a short excerpt from a motion picture for criticism or comment.

The committee's attention has been directed to the special problems involved in the reception of instructional television programs in remote areas of the country. In certain areas it is currently impossible to transmit such programs by any means other than communications satellites. A particular difficulty exists when such transmissions extend over several time zones within the same state, such as in Alaska. Unless individual schools in such states may make an off-air recording of such transmissions, the programs may not be received by the students during the school's daily schedule. The committee believes that the making by a school located in such a remote area of an off-the-air recording of an instructional television transmission for the purpose of a delayed viewing of the program by students for the same school constitutes a "fair use." The committee does not intend to suggest however, that off-the-air recording for convenience would under any circumstances, be considered "fair use." To meet the requirement of temporary use the school may retain the recording for only a limited period of time after the broadcast.

Another special instance illustrating the application of the fair use doctrine pertains to the making of copies or phonorecords of works in the special forms needed for the use of blind persons. These special forms, such as copies in braille and phonorecords of oral readings (talking books), are not usually made by the publishers for commercial distribution. For the most part, such copies and phonorecords are made by the Library of Congress' Division for the Blind and Physically Handicapped with permission obtained from the copyright owners, and are circulated to blind persons through regional libraries covering the nation. In addition, such copies and phonorecords are made locally by individual volunteers for the use of blind persons in their communities, and the Library of Congress conducts a program for training such volunteers. While the making of multiple copies or phonorecords of a work for general circulation requires the permission of the copyright owner, the making of a single copy or phonorecord by an individual as a free service for a blind person would properly be considered a fair use under section 107.

A problem of particular urgency is that of preserving for posterity prints of motion pictures made before 1942. Aside from the deplorable fact that in a great many cases the only existing copy of a film has been deliberately destroyed, those that remain are in immediate danger of disintegration; they were printed on film stock with a nitrate base that will inevitably decompose in time. The efforts of the Library of Congress, the American Film Institute, and other organizations to rescue and preserve this irreplaceable contribution to our cultural life are to be applauded, and the making of duplicate copies for purposes of archival preservation certainly falls within the scope of "fair use". ·

When a copyrighted work contains unfair, inaccurate, or derogatory information concerning an individual or institution, such individual or institution may copy and reproduce such parts of the work as are necessary to permit understandable comment on the statements made in the work.

During the consideration of the revision bill in the 94th Congress it was proposed that independent newsletters, as distinguished from house organs and publicity or advertising publications, be given separate treatment. It is argued that newsletters are particularly vulnerable to mass photocopying, and that most newsletters have fairly modest circulations. Whether the copying of portions of a newsletter is an act of infringement or a fair use must be judged by the general provisions of this legislation. However, the copying of even a short portion of a newsletter may have a significant impact on the commercial market for the work.

The committee has examined the use of excerpts from copyrighted works in the art work of calligraphers. The committee believes that a single copy reproduction of an excerpt from a copyrighted work by a calligrapher for a single client does not represent an infringement of copyright. Likewise, a single reproduction of excerpts from a copyrighted work by a student calligrapher or teacher in a learning situation would be a fair use of the copyrighted work.

The Register of Copyrights has recommended that the committee report describe the relationship between this section and the provisions of section 108 relating to reproduction by libraries and archives. The doctrine of fair use applies to library photocopying, and nothing contained in section 108 "in any way affects the right of fair use." No provision of section 108 is intended to take away any rights existing under the fair use doctrine. To the contrary, section 108 authorizes certain photocopying practices which may not qualify as a fair use.

The criteria of fair use are necessarily set forth in general terms. In the application of the criteria of fair use to specific photocopying practices of libraries, it is the intent of this legislation to provide an appropriate balancing of the rights of creators, and the needs of users.

Appendix C

Committee Reports on Section 108, Reproduction by Libraries and Archives

(1) House Committee Report on the 1976 Copyright Bill (House Committee on the Judiciary, House Report No. 94-1476 to accompany S. 22, 94th Cong., 2d Sess., September 3, 1976, pp. 74–79)

Notwithstanding the exclusive rights of the owners of copyright, section 108 provides that under certain conditions it is not an infringement of copyright for a library or archives, or any of its employees acting within the scope of their employment, to reproduce or distribute not more than one copy or phonorecord of a work, provided (1) the reproduction or distribution is made without any purpose of direct or indirect commercial advantage and (2) the collections of the library or archives are open to the public or available not only to researchers affiliated with the library or archives, but also to other persons doing research in a specialized field, and (3) the reproduction or distribution of the work includes a notice of copyright.

Under this provision, a purely commercial enterprise could not establish a collection of copyrighted works, call itself a library or archive, and engage in for-profit reproduction and distribution of photocopies. Similarly, it would not be possible for a non-profit institution, by means of contractual arrangements with a commercial copying enterprise, to authorize the enterprise to carry out copying and distribution functions that would be exempt if conducted by the nonprofit institution itself.

The reference to "indirect commercial advantage" has raised questions as to the status of photocopying done by or for libraries or archival collections within industrial, profitmaking, or proprietary institutions (such as the research and development departments of chemical, pharmaceutical, automobile, and oil corporations, the library of a propritary hospital, the collections owned by a law or medical partnership, etc.).

There is a direct interrelationship between this problem and the prohibitions against "multiple" and "systematic" photocopying in section 108(g) (1) and (2). Under section 108, a library in a profitmaking organization would not be authorized to:

(a) use a single subscription or copy to supply its employees with multiple copies of material relevant to their work; or

(b) use a single subscription or copy to supply its employees, on request, with single copies of material relevant to their work, where the arrangement is "systematic" in the sense of deliberately substituting photocopying for subscription or purchase; or

(c) use "interlibrary loan" arrangements for obtaining photocopies in such aggregate quantities as to substitute for subscriptions or purchase of material needed by employees in their work.

Moreover, a library in a profit-making organization could not evade these obligations by installing reproducing equipment on its premises for unsupervised use by the organization's staff.

Isolated, spontaneous making of single photocopies by a library in a for-profit organization, without any systematic effort to substitute photocopying for subscriptions or purchases, would be covered by section 108, even though the copies are furnished to the employees of the organization for use in their work. Similarly, for-profit libraries could participate in interlibrary arrangements for exchange of photocopies, as long as the reproduction or distribution was not "systematic." These activities, by themselves, would ordinarily not be considered "for direct or indirect commercial advantage;" since the "advantage" referred to in this clause must attach to the immediate commercial motivation behind the reproduction or distribution itself, rather than to the ultimate profit-making motivation behind the enterprise in which the library is located. On the other hand, section 108 would not excuse reproduction or distribution if there were a commercial motive behind the actual making or distributing of the copies, if multiple copies were made or distributed, or if the photocopying activities were "systematic" in the sense that their aim was to substitute for subscriptions or purchases.

The rights of reproduction and distribution under section 108 apply in the following circumstances:

Archival Reproduction

Subsection (b) authorizes the reproduction and distribution of a copy or phonorecord of an unpublished work duplicated in facsimile form solely for purposes of preservation and security, or for deposit for research use in another library or archives, if the copy or phonorecord reproduced is currently in the collections of the first library or archives.

Only unpublished works could be reproduced under this exemption, but the right would extend to any type of work, including photographs, motion pictures and sound recordings. Under this exemption, for example, a repository could make photocopies of manuscripts by microfilm or electrostatic process, but could not reproduce the work in "machine-readable" language for storage in an information system.

Replacement of Damaged Copy

Subsection (c) authorizes the reproduction of a published work duplicated in facsimile form solely for the purpose of replacement of a copy or phonorecord that is damaged, deteriorating, lost or stolen, if the library or archives has, after a reasonable effort, determined that an unused replacement cannot be obtained at a fair price. The scope and nature of a reasonable investigation to determine that an unused replacement cannot be obtained will vary according to the circumstances of a particular situation. It will always require recourse to commonly-known trade sources in the United States, and in the normal situation also to the publisher or other copyright owner (if such owner can be located at the address listed in the copyright registration), or an authorized reproducing service.

Articles and Small Excerpts

Subsection (d) authorizes the reproduction and distribution of a copy of not more than one article or other contribution to a copyrighted collection or periodical issue, or of a copy or phonorecord of a small part of any other copyrighted work. The copy or phonorecord may be made by the library where the user makes his request or by another library pursuant to an interlibrary loan. It is further required that the copy become the property of the user, that the library or archives have no notice that the copy would be used for any purposes other than private study, scholarship or research, and that the library or archives display prominently at the place where reproduction requests are accepted, and includes in its order form, a warning of copyright in accordance with requirements that the Register of Copyrights shall prescribe by regulation.

Out-of-Print Works

Subsection (e) authorizes the reproduction and distribution of a copy or phonorecord of an entire work under certain circumstances, if it has been established that a copy cannot be obtained at a fair price. The copy may be made by the library where the user makes his request or by another library pursuant to an interlibrary loan. The scope and nature of a reasonable investigation to determine that an unused copy cannot be obtained will vary according to the circumstances of a particular situation. It will always require recourse to commonly-known trade sources in the United States, and in the normal situation also to the publisher or other copyright owner (if the owner can be located at the address listed in the copyright registration), or an authorized reproducing service. It is further required that the copy become the property of the user, that the library or archives have no notice that the copy would be used for any purpose other than private study, scholarship, or research, and that the library or archives display prominently at the place where reproduction requests are accepted, and include on its order form, a warning of copyright in accordance with requirements that the Register of Copyrights shall prescribe by regulation.

General Exemptions

Clause (1) of subsection (f) specifically exempts a library or archives or its employees from liability for the unsupervised use of reproducing equipment located on its premises, provided that the reproducing equipment displays a notice that the making of a copy may be subject to the copyright law. Clause (2) of subsection (f) makes clear that this exemption of the library or archives does not extend to the person using such equipment or requesting such copy if the use exceeds fair use. Insofar as such person is concerned the copy or phonorecord made is not considered "lawfully" made for purposes of sections 109, 110 or other provisions of the title.

Clause (3) provides that nothing in section 108 is intended to limit the reproduction and distribution by lending of a limited number of copies and excerpts of an audiovisual news program. This exemption is intended to apply to the daily newscasts of the national television networks, which report the major events of the day. It does not apply to documentary (except documentary programs involving news reporting as that term is used in section 107), magazine-format or other public affairs broadcasts dealing with subjects of general interest to the viewing public.

The clause was first added to the revision bill in 1974 by the adoption of an amendment proposed by Senator Baker. It is intended to permit libraries and archives, subject to the general conditions of this section, to make off-the-air videotape recordings of daily network newscasts for limited distribution to scholars and researchers for use in research purposes. As such, it is an adjunct to the American Television and Radio Archive established in Section 113 of the Act which will be the principal repository for television broadcast material, including news broadcasts. The inclusion of language indicating that such material may only be distributed by lending by the library or archive is intended to preclude performance, copying, or sale, whether or not for profit, by the recipient of a copy of a television broadcast taped off-the-air pursuant to this clause.

Clause (4), in addition to asserting that nothing contained in section 108 "affects the right of fair use as provided by section 107," also provides that the right of reproduction granted by this section does not override any contractual arrangements assumed by a library or archives when it obtained a work for its collections. For example, if there is an express contractual prohibition against reproduction for any purpose, this legislation shall not be construed as justifying a violation of the contract. This clause is intended to encompass the situation where an individual makes papers, manuscripts or other works available to a library with the understanding that they will not be reproduced.

It is the intent of this legislation that a subsequent unlawful use by a user of a copy of phonorecord of a work lawfully made by a library, shall not make the library liable for such improper use.

Multiple Copies and Systematic Reproduction

Subsection (g) provides that the rights granted by this section extend only to the "isolated and unrelated reproduction of a single copy or phonorecord of the same material on separate occasions." However, this section does not authorize the related or concerted reproduction of multiple copies or phonorecords of the same material, whether made on one occasion or over a period of time, and whether intended for aggregate use by one individual or for separate use by the individual members of a group.

With respect to material described in subsection (d)—articles or other contributions to periodicals or collections, and small parts of other copyrighted works—subsection (g) (2) provides that the exemptions of section 108 do not apply if the library or archive engages in "systematic reproduction or distribution of single or multiple copies or phonorecords." This provision in S. 22 provoked a storm of controversy, centering around the extent to which the restrictions on "systematic" activities would prevent the continuation and development of interlibrary networks and other arrangements involving the exchange of photocopies. After thorough consideration, the Committee amended section 108(g)(2) to add the following proviso:

Provided, that nothing in this clause prevents a library or archives from participating in interlibrary arrangements that do not have, as their purpose or effect, that the library or archives receiving such copies or phonorecords for distribution does so in such aggregate quantities as to substitute for a subscription to or purchase of such work.

In addition, the Committee added a new subsection (i) to section 108, requiring the Register of Copyrights, five years from the effective date of the new Act and at five-year intervals thereafter, to report to Congress upon "the extent to which this section has achieved the intended statutory balancing of the rights of creators, and the needs of users," and to make appropriate legislative or other recommendations. As noted in connection with section 107, the Committee also amended section 504(c) in a way that would insulate librarians from unwarranted liability for copyright infringement; this amendment is discussed below.

The key phrases in the Committee's amendment of section 108(g)(2) are "aggregate quantities" and "substitute for a subscription to or purchase of" a work. To be implemented effectively in practice, these provisions will require the development and implementation of more-or-less specific guidelines establishing criteria to govern various situations.

The National Commission on New Technological Uses of Copyrighted Works (CONTU) offered to provide good offices in helping to develop these guidelines. This offer was accepted and, although the final text of guidelines has not yet been achieved, the Committee has reason to hope that, within the next month, some agreement can be reached on an initial set of guidelines covering practices under section 108(g)(2).

Works Excluded

Subsection (h) provides that the rights of reproduction and distribution under this section do not apply to a musical work, a pictorial, graphic or sculptural work, or a motion picture or other audiovisual work other than "an audiovisual work dealing with news." The latter term is intended as the equivalent in meaning of the phrase "audiovisual news program" in section 108(f)(3). The exclusions under subsection (h) do not apply to archival reproduction under subsection (b), to replacement of damaged or lost copies or phonorecords under subsection (c), or to "pictorial or graphic works published as illustrations, diagrams, or similar adjuncts to works of which copies are reproduced or distributed in accordance with subsections (d) and (e)."

Although subsection (h) generally removes musical, graphic, and audiovisual works from the specific exemptions of section 108, it is important to recognize that the doctrine of fair use under section 107 remains fully applicable to the photocopying or other reproduction of such works. In the case of music, for example, it would be fair use for a scholar doing musicological research to have a library supply a copy of a portion of a score or to reproduce portions of a phonorecord of a work. Nothing in section 108 impairs the applicability of the fair use doctrine to a wide variety of situations involving photocopying or other reproduction by a library of copyrighted material in its collections, where the user requests the reproduction for legitimate scholarly or research purposes.

(2) Senate Committee Report on the 1976 Copyright Bill (Senate Committee on the Judiciary, Senate Report No. 94-473 to accompany S. 22, 94th Cong., 1st Sess., November 20, 1975, pp. 67–71)

Notwithstanding the exclusive rights of the owners of copyright, section 108 provides that under certain conditions it is not an infringement of copyright for a library or archives, or any of their employees acting within the scope of their employment, to reproduce or distribute not more than one copy of phonorecord of a work provided (1) the reproduction or distribution is made without any purpose of direct or indirect commercial advantage and (2) the collections of the library or archives are open to the public or available not only to researchers affiliated with the library or archives, but also to other persons doing re search in a specialized field, and (3) the reproduction or distribution of the work includes a notice of copyright.

The limitation of section 108 to reproduction and distribution by libraries and archives "without any purpose of direct or indirect commercial advantage" is intended to preclude a library or archives in a profit-making organization from providing photocopies of copyrighted materials to employees engaged in furtherance of the organization's commercial enterprise, unless such copying qualifies as a fair use, or the organization has obtained the necessary copyright licenses. A commercial organization should purchase the number of copies of a work that it requires, or obtain the consent of the copyright owner to the making of the photocopies.

The rights of reproduction and distribution under section 108 apply in the following circumstances:

Archival Reproduction

Subsection (b) authorizes the reproduction and distribution of a copy or phonorecord of an unpublished work duplicated in facsimile form solely for purposes of preservation and security, or for deposit for research use in another library or archives, if the copy or phonorecord reproduced is currently in the collections of the first library or archives.

Only unpublished works could be reproduced under this exemption, but the right would extend to any type of work, including photographs, motion pictures and sound recordings. Under this exemption, for example, a repository could make photocopies of manuscripts by microfilm or electrostatic process, but could not reproduce the work in "machine-readable" language for storage in an information system.

Replacement of Damaged Copy

Subsection (c) authorizes the reproduction of a published work duplicated in facsimile form solely for the purpose of replacement of a copy or phonorecord that is damaged, deteriorating, lost, or stolen, if the library or archives has, after a reasonable effort, determined that an unused replacement cannot be obtained at a fair price. The scope and nature of a reasonable investigation to determine that an unused replacement cannot be obtained will vary according to the circumstances of a particular situation. It will always require recourse to commonly-known trade sources in the United States, and in the normal situation also to the published or other copyright owner (if such owner can be located at the address listed in the copyright registration), or an authorized reproducing service.

Articles and Small Excerpts

Subsection (d) authorizes the reproduction and distribution of a copy of not more than one article or other contribution to a copyrighted collection of a periodical or copy or phonorecord of a small part of any other copyrighted work. The copy may be made by the library where the user makes his request or by another library pursuant to an inter-library loan. It is further required that the copy become the property of the user, that the library or archives have no notice that the copy would be used for any purposes other than private study, scholarship or research, and that the library or archives display prominently at the place whether reproduction requests are accepted. And includes in its order form, a warning of copyright in accordance with requirements that the Register of Copyrights shall prescribe by regulation.

Out-of-Print Works

Subsection (c) authorizes the reproduction and distribution of a copy of a work, with certain exceptions, at the request of the user of the collection if the user has established that an unused copy cannot be obtained at a fair price. The copy may be made by the library where the user makes his request or by another library pursuant to an inter-library loan. The scope and nature of a reasonable investigation to determine that an unused copy cannot be obtained will vary according to the circumstances of a particular situation. It will always require recourse to commonly-known trade sources in the United States, and in the normal situation also to the publisher or other copyright owner (if the owner can be located at the address listed in the copyright registration), or an authorized reproducing service. It is further required that the copy become the property of the user, that the library or archives have no notice that the copy would be used for any purpose other than private study, scholarship, or research, and that the library or archives display prominently at the place where reproduction requests are accepted, and include on its order form, a warning of copyright in accordance with requirements that the Register of Copyright shall prescribe by regulation.

General Exemptions

Clause (1) of subsection (f) specifically exempts a library or archives or their employees from such liability provided that the reproducing equipment displays a notice that the making of a copy may be subject to the copyright law. Clause (2) of subsection (f) makes clear that this exemption of the library or archives does not extend to the person using such equipment or requesting such copy if the use exceeds fair use. Insofar as such person is concerned the copy made is not considered "lawfully" made for purposes of sections 109, 110 or other provisions of the title. Clause (3) in addition to asserting that nothing contained in section 108 "affects the right of fair use as provided by section 107," also provides that the right of reproduction granted by this section does not override any contractual arrangements assumed by a library or archives when it obtained a work for its collections. For example, if there is an express contractual prohibition against reproduction for any purpose, this legislation shall not be construed as justifying a violation of the contract. This clause is intended to encompass the situation where an individual makes papers, manuscripts or other works available to a library with the understanding that they will not be reproduced.

Clause (4) provides that nothing in section 108 is intended to limit the reproduction and distribution of a limited number of copies and excerpts of an audiovisual news program.

This clause was first added to the revision bill last year by the adoption of an amendment proposed by Senator Baker. It is intended to permit libraries and archives, subject to the general conditions of this section, to make off-the-air videotape recordings of television news programs. Despite the importance of preserving television news, the United States currently has no institution performing this function on a systematic basis.

The purpose of the clause is to prevent the copyright law from precluding such operations as the Vanderbilt University Television News Archive, which makes videotape recordings of television news programs, prepares indexes of the contents, and leases copies of complete broadcasts or compilations of coverage of specified subjects for limited periods upon request from scholars and researchers.

Because of the important copyright policy issues inherent in this issue, the exemption has been narrowly drafted. The Register of Copyrights in 1974 advised that the language of this clause was technically appropriate for its purpose and not "broader than is necessary to validate the Vanderbilt operation."

The Copyright Office recommended that if the Congress desires a news videotape exemption it should be incorporated in section 108. The Copyright Office stated that the inclusion of such a clause in section 108 would be adequate "to enable the Vanderbilt operation to continue."

It is the intent of this legislation that a subsequent unlawful use by a user of a copy of a work lawfully made by a library, shall not make the library liable for such improper use.

Multiple Copies and Systematic Reproduction

Subsection (g) provides that the rights granted by this section extend only to the "isolated and unrelated reproduction of a single copy", but this section does not authorize the related or concerted reproduction of multiple copies of the same material whether made on one occasion or over a period of time, and whether intended for aggregate use by one individual or for separate use by the individual members of a group. For example, if a college professor instructs his class to read an article from a copyrighted journal, the school library would not be permitted, under subsection (g), to reproduce copies of the article for the members of the class.

Subsection (g) also provides that section 108 does not authorize the systematic reproduction or distribution of copies or phonorecords of articles or other contributions to copyrighted collections or periodicals or of small parts of other copyrighted works whether or not multiple copies are reproduced or distributed. Systematic reproduction or distribution occurs when a library makes copies of such materials available to other libraries or to groups of users under formal or informal arrangements whose purpose or effect is to have the reproducing library serve as their source of such material. Such systematic reproduction and distribution, as distinguished from isolated and unrelated reproduction or distribution, may substitute the copies reproduced by the source library for subscriptions or reprints or other copies which the receiving libraries or users might otherwise have purchased for themselves, from the publisher or the licensed reproducing agencies.

While it is not possible to formulate specific definitions of "systematic copying", the following examples serve to illustrate some of the copying prohibited by subsection (g).

(1) A library with a collection of journals in biology informs other libraries with similar collections that it will maintain and build its own collection and will make copies of articles from these journals available to them and their patrons on request. Accordingly, the other libraries discontinue or refrain from purchasing subscriptions to these journals and fulfill their patrons' requests for articles by obtaining photocopies from the source library.

(2) A research center employing a number of scientists and technicians subscribes to one or two copies of needed periodicals. By reproducing photocopies of articles the center is able to make the material in these periodicals available to its staff in the same manner which otherwise would have required multiple subscriptions.

(3) Several branches of a library system agree that one branch will subscribe to particular journals in lieu of each branch purchasing its own subscriptions, and the one subscribing branch will reproduce copies of articles from the publication for users of the other branches.

The committee believes that section 108 provides an appropriate statutory balancing of the rights of creators, and the needs of users. However, neither a statute nor legislative history can specify precisely which library photocopying practices constitute the making of "single copies" as distinguished from "systematic reproduction". Isolated single spontaneous requests must be distinguished from "systematic reproduction". The photocopying needs of such operations as multicounty regional systems must be met. The committee therefore recommends that representatives of authors, book and periodical publishers and other owners of copyrighted material meet with the library community to formulate photocopying guidelines to assist library patrons and employees. Concerning library photocopying practices not authorized by this legislation, the committee recommends that workable clearance and licensing procedures be developed.

It is still uncertain how far a library may go under the Copyright Act of 1909 in supplying a photocopy of copyrighted material in its collection. The recent case of *The Williams and Wilkins Company* v. *The United States* failed to significantly illuminate the application of the fair use doctrine to library photocopying practices. Indeed, the opinion of the Court of Claims said the Court was engaged in "a 'holding operation' in the interim period before Congress enacted its preferred solution."

While the several opinions in the *Wilkins* case have given the Congress little guidance as to the current state of the law on fair use, these opinions provide additional support for the balanced resolution of the photocopying issue adopted by the Senate last year in S. 1361 and preserved in section 108 of this legislation. As the Court of Claims opinion succinctly stated "there is much to be said on all sides."

In adopting these provisions on library photocopying, the committee is aware that through such programs as those of the National Commission on Libraries and Information Science there will be a significant evolution in the functioning and services of libraries. To consider the possible need for changes in copyright law and procedures as a result of new technology, a National Commission on New Technological Uses of Copyrighted Works has been established (Public Law 93–573).

Works Excluded

Subsection (h) provides that the rights of reproduction and distribution under this section do not apply to a musical work, a pictorial, graphic or sculptural work, or a motion picture or other audio-visual work. Such limitation does not apply to archival reproduction and replacement of a damaged copy.

(3) Conference Report on the 1976 Copyright Bill (Conference Report, House Report No. 94-1733 to accompany S. 22, 94th Cong., 2d Sess., September 29, 1976, pp. 70–74)

Senate Bill

Section 108 of the Senate bill dealt with a variety of situations involving photocopying and other forms of reproduction by libraries and archives. It specified the conditions under which single copies of copyrighted material can be noncommercially reproduced and distributed, but made clear that the privileges of a library or archives under the section do not apply where the reproduction or distribution is of multiple copies or is "systematic." Under subsection (f), the section was not to be construed as limiting the reproduction and distribution, by a library or archive meeting the basic criteria of the section, of a limited number of copies and excerpts of an audiovisual news program.

House Bill

The House bill amended section 108 to make clear that, in cases involving interlibrary arrangements for the exchange of photocopies, the activity would not be considered "systematic" as long as the library or archives receiving the reproductions for distribution does not do so in such aggregate quantities as to substitute for a subscription to or purchase of the work. A new subsection (i) directed the Register of Copyrights, by the end of 1982 and at five-year intervals thereafter, to report on the practical success of the section in balancing the various interests, and to make recommendations for any needed changes. With respect to audiovisual news programs, the House bill limited the scope of the distribution privilege confirmed by section 108(f)(3) to cases where the distribution takes the form of a loan.

Conference Substitute

The conference substitute adopts the provisions of section 108 as amended by the House bill. In doing so, the conferees have noted two letters dated September 22, 1976, sent respectively to John L. McClellan, Chairman of the Senate Judiciary Subcommittee on Patents, Trademarks, and Copyrights, and to Robert W. Kastenmeier, Chairman of the House Judiciary Subcommittee on Courts, Civil Liberties, and the Administration of Justice. The letters, from the Chairman of the National Commission on New Technological Uses of Copyrighted Works (CONTU), Stanley H. Fuld, transmitted a document consisting of "guidelines interpreting the provision in subsection 108(g)(2) of S. 22, as approved by the House Committee on the Judiciary." Chairman Fuld's letters explain that, following lengthy consultations with the parties concerned, the Commission adopted these guidelines as fair and workable and with the hope that the conferees on S. 22 may find that they merit inclusion in the conference report. The letters add that, although time did not permit securing signatures of the representatives of the principal library organizations or of the organizations representing publishers and authors on these guidelines, the Commission had received oral assurances from these representatives that the guidelines are acceptable to their organizations.

The conference committee understands that the guidelines are not intended as, and cannot be considered, explicit rules or directions governing any and all cases, now or in the future. It is recognized that their purpose is to provide guidance in the most commonly-encountered interlibrary photocopying situations, that they are not intended to be limiting or determinative in themselves or with respect to other situations, and that they deal with an evolving situation that will undoubtedly require their continuous reevaluation and adjustment. With these qualifications, the conference committee agrees that the guidelines are a reasonable interpretation of the proviso of section 108(g)(2) in the most common situations to which they apply today.

The text of the guidelines follows:

PHOTOCOPYING—INTERLIBRARY ARRANGEMENTS

INTRODUCTION

Subsection 108(g)(2) of the bill deals, among other things, with limits on interlibrary arrangements for photocopying. It prohibits systematic photocopying of copyrighted materials but permits interlibrary arrangements "that do not have, as their purpose or effect, that the library or archives receiving such copies or phonorecords for distribution does so in such aggregate quantities as to substitute for a subscription to or purchase of such work."

The National Commission on New Technological Uses of Copyrighted Works offered its good offices to the House and Senate subcommittees in bringing the interested parties together to see if agreement could be reached on what a realistic definition would be of "such aggregate quantities." The Commission consulted with the parties and suggested the interpretation which follows, on which there has been substantial agreement by the principal library, publisher, and author organizations. The Commission considers the guidelines which follow to be a workable and fair interpretation of the intent of the proviso portion of subsection 108(g)(2).

These guidelines are intended to provide guidance in the application of section 108 to the most frequently encountered interlibrary case: a library's obtaining from another library, in lieu of interlibrary loan, copies of articles from relatively recent issues of periodicals—those published within five years prior to the date of the request. The guidelines do not specify what aggregate quantity of copies of an article or articles published in a periodical, the issue date of which is more than five years prior to the date when the request for the copy thereof is made, constitutes a substitute for a subscription to such periodical. The meaning of the proviso to subsection 108(g)(2) in such case is left to future interpretation.

The point has been made that the present practice on interlibrary loans and use of photocopies in lieu of loans may be supplemented or even largely replaced by a system in which one or more agencies or institutions, public or private, exist for the specific purpose of providing a central source for photocopies. Of course, these guidelines would not apply to such a situation.

GUIDELINES FOR THE PROVISO OF SUBSECTION 108(G)(2)

1. As used in the proviso of subsection 108(g)(2), the words ". . . such aggregate quantities as to substitute for a subscription to or purchase of such work" shall mean:

(a) with respect to any given periodical (as opposed to any given issue of a periodical), filled requests of a library or archives (a "requesting entity") within any calendar year for a total of six or more copies of an article or articles published in such periodical within five years prior to the date of the request. These guidelines specifically shall not apply, directly or indirectly, to any request of a requesting entity for a copy or copies of an article or articles published in any issue of a periodical, the publication date of which is more than five years prior to the date when the request is made. These guidelines do not define the meaning, with respect to such a request, of ". . . such aggregate quantities as to substitute for a subscription to [such periodical]".

(b) With respect to any other material described in subsection 108(d), (including fiction and poetry), filled requests of a requesting entity within any calendar year for a total of six or more copies or phonorecords of or from any given work (including a collective work) during the entire period when such material shall be protected by copyright.

2. In the event that a requesting entity—

(a) shall have in force or shall have entered an order for a subscription to a periodical, or

(b) has within its collection, or shall have entered an order for, a copy or phonorecord of any other copyrighted work,

material from either category of which it desires to obtain by copy from another library or archives (the "supplying entity"), because the material to be copied is not reasonably available for use by the requesting entity itself, then the fulfillment of such request shall be treated as though the requesting entity made such copy from its own collection. A library or archives may request a copy or phonorecord from a supplying entity only under those circumstances where the requesting entity would have been able, under the other provisions of section 108, to supply such copy from materials in its own collection.

3. No request for a copy or phonorecord of any material to which these guidelines apply may be fulfilled by the supplying entity unless such request is accompanied by a representation by the requesting entity that the request was made in conformity with these guidelines.

4. The requesting entity shall maintain records of all requests made by it for copies or phonorecords of any materials to which these guidelines apply and shall maintain records of the fulfillment of such requests, which records shall be retained until the end of the third complete calendar year after the end of the calendar year in which the respective request shall have been made.

5. As part of the review provided for in subsection 108(i), these guidelines shall be reviewed not later than five years from the effective date of this bill.

The conference committee is aware that an issue has arisen as to the meaning of the phrase "audiovisual news program" in section 108(f)(3). The conferees believe that, under the provision as adopted in the conference substitute, a library or archives qualifying under section 108(a) would be free, without regard to the archival activities of the Library of Congress or any other organization, to reproduce, on videotape or any other medium of fixation or reproduction, local, regional, or network newscasts, interviews concerning current news events, and on-the-spot coverage of news events, and to distribute a limited number of reproductions of such a program on a loan basis.

Another point of interpretation involves the meaning of "indirect commercial advantage," as used in section 108(a)(1), in the case of libraries or archival collections within industrial, profit-making, or proprietary institutions. As long as the library or archives meets the criteria in section 108(a) and the other requirements of the section, including the prohibitions against multiple and systematic copying in subsection (g), the conferees consider that the isolated, spontaneous making of single photocopies by a library or archives in a for-profit organization without any commercial motivation, or participation by such a library or archives in interlibrary arrangements, would come within the scope of section 108.

Appendix D

1965 Supplementary Report of the Register of Copyrights on the General Revision of the U.S. Copyright Law, 89th Cong., 1st Sess., Copyright Law Revision, Part 6 (House Judiciary Comm. Print, 1965), pp. 25–28

C. Fair Use

Although it is not mentioned in the present statute, the doctrine of fair use, as it has been developed in a long line of court decisions, is probably the most significant and widely applicable of the limitations on the copyright owner's exclusive rights. The 1961 *Report* described the general scope of the doctrine and gave a number of examples of cases where the concept would be relevant. It was acknowledged, however, that fair use "eludes precise definition" and that, because of the number and variety of situations in which fair use could be involved, "it would be difficult to prescribe precise rules suitable for all occasions."

The *Report* concluded that "the doctrine of fair use is such an important limitation on the rights of copyright owners, and occasions to apply that doctrine arise so frequently, that we believe the statute should mention it and indicate its general scope." As a special aspect of fair use the *Report* also discussed the problem of photocopying by libraries for research purposes, and recommended that the statute include provisions permitting a library to supply single photocopies under specified conditions and within certain limits.

These recommendations were carried over into the preliminary draft of 1963. Section 6, dealing with the general concept of fair use, provided:

All of the exclusive rights specified in section 5 shall be limited by the privilege of making fair use of a copyrighted work. In determining whether, under the circumstances in any particular case, the use of a copyrighted work constitutes a fair use rather than an infringement of copyright, the following factors, among others, shall be considered: (a) the purpose and character of the use, (b) the nature of the copyrighted work, (c) the amount and substantiality of the material used in relation to the copyrighted work as a whole, and (d) the effect of the use upon the potential value of the copyrighted work.

Section 7 was a rather elaborate provision which, in general, would have permitted libraries to supply a single photocopy of one article (or other contribution or excerpt) from a copyrighted work, or a single photocopy of an entire work if it were out of print.

The language of section 6 met with a certain amount of favor, but section 7 was strenuously opposed on all sides. Author and publisher groups attacked section 7 as opening the door to wholesale and unrestrained copying by libraries which, as reproduction equipment improves, could supplant the copies offered for sale by publishers and undercut the author's main source of remuneration. Library groups were equally vehement in opposition to the proposals, which they argued would curtail established services and prevent the free utilization of new devices in the interests of research and scholarship.

In a way the comments on section 7 of the preliminary draft represented an interesting case study. Opposition to the provision was equally strong on both sides but for exactly opposite reasons, with one side arguing that the provision would permit things that are illegal now and the other side maintaining that it would prevent things that are legal now. Both agreed on one thing: that the section should be dropped entirely. We also became convinced that the provision would be a mistake in any event. At the present time the practices, techniques, and devices for reproducing visual images and sound and for "storing" and "retrieving" information are in such a stage of rapid evolution that any specific statutory provision would be likely to prove inadequate, if not unfair or dangerous, in the not too distant future. As important as it is, library copying is only one aspect of the much larger problem of changing technology, and we feel the statute should deal with it in terms of broad fundamental concepts that can be adapted to future developments.

The decision to drop any provision on photocopying tended to increase the importance attached to including a general section on fair use in the statute. Thus, in the 1964 bill, further language was added to section 6 in an attempt to clarify the scope of the doctrine of fair use but without freezing or delimiting its application to new uses:

Notwithstanding the provisions of section 5, the fair use of a copyrighted work to the extent reasonably necessary or incidental to a legitimate purpose such as criticism, comment, news reporting, teaching, scholarship, or research is not an infringement of copyright. In determining whether the use made of a work in any particular case is a fair use, the factors to be considered shall include:

(1) the purpose and character of the use;

(2) the nature of the copyrighted work;

(3) the amount and substantiality of the portion used in relation to the copyrighted work as a whole; and

(4) the effect of the use upon the potential market for or value of the copyrighted work.

This language elicited a large body of comments, most of them critical. Without reviewing the arguments in detail, it can be said in general that the author-publisher groups expressed fears that specific mention of uses such as "teaching, scholarship, or research" could be taken to imply that any use even remotely connected with these activities would be a "fair use." On the other side, serious objections were raised to the use of qualifying language such as "to the extent reasonably necessary or incidental to a legitimate purpose" and "the amount and substantiality of the portion used . . . "

In addition to opposing this language as unduly restrictive, a group of educational organizations urged that the bill adopt a new provision which would specify a number of activities involved in teaching and scholarship as completely exempt from copyright control. In broad terms, and with certain exceptions, the proposal as it evolved would permit any teacher or other person or organization engaged in non-profit educational activities to make a single copy or record of an entire work, or a reasonable number of copies of "excerpts or quotations," for use in connection with those activities. It was argued that these privileges are a necessary part of good teaching, and that it is unjustifiable to burden educators with the need to buy copies for limited use or to obtain advance clearances and pay royalties for making copies. These proposals were opposed very strongly by authors, publishers, and other copyright owners on the ground that in the short run the reproduction of copies under this proposal would severely diminish the market for their works, and that the ultimate result would be to destroy the economic incentive for the creation and publication of the very works on which education depends for its existence. It was suggested that a clearinghouse for educational materials, through which it would be possible to avoid problems of clearances, is a practical possibility for the near future

For reasons we have already discussed at some length, we do not favor sweeping, across-the-board exemptions from the author's exclusive rights unless an overriding public need can be conclusively demonstrated. There is hardly any public need today that is more urgent than education, but we are convinced that this need would be ill-served if educators, by making copies of the materials they need, cut off a large part of the revenue to authors and publishers that induces the creation and publication of those materials. We believe that a statutory recognition of fair use would be sufficient to serve the reasonable needs of education with respect to the copying of short extracts from copyrighted works, and that the problem of obtaining clearances for copying larger portions of entire works could best be solved through a clearing house arrangement worked out between the educational groups and the author-publisher interests.

Since it appeared impossible to reach agreement on a general statement expressing the scope of the fair use doctrine, and since in any event the doctrine emerges from a body of judicial precedent and not from the statute, we decided with some regret to reduce the fair use section to its barest essentials. Section 107 of the 1965 bill therefore provides:

> Notwithstanding the provisions of section 106, the fair use of a copyrighted work is not an infringement of copyright.

We believe that, even in this form, the provision serves a real purpose and should be incorporated in the statute.

The author-publisher interests have suggested that fair use should be treated as a defense, with the statute placing the burden of proof on the user. The educational group has urged just the opposite, that the statute should provide that any nonprofit use for educational purposes is presumed to be a fair use, with the copyright owner having the burden of proving otherwise. We believe it would be undesirable to adopt a special rule placing the burden of proof on one side or the other. When the facts as to what use was made of the work have been presented, the issue as to whether it is a "fair use" is a question of law. Statutory presumptions or burden-of-proof provisions could work a radical change in the meaning and effect of the doctrine of fair use. The intention of section 107 is to give statutory affirmation to the present judicial doctrine, not to change it.

Appendix E

1935 "Gentlemen's Agreement"

The Joint Committee on Materials for Research and the Board of Directors of the National Association of Book Publishers, after conferring on the problem of conscientious observance of copyright that faces research libraries in connection with the growing use of photographic methods of reproduction, have agreed upon the following statement:

A library, archives office, museum, or similar institution owning books or periodical volumes in which copyright still subsists may make and deliver a single photographic reproduction or reduction of a part thereof to a scholar representing in writing that he desires such reproduction in lieu of loan of such publication or in place of manual transcription and solely for the purposes of research; provided

(1) That the person receiving it is given due notice in writing that he is not exempt from liability to the copyright proprietor for any infringement of copyright by misuse of the reproduction constituting an infringement under the copyright law;

(2) That such reproduction is made and furnished without profit to itself by the institution making it.

The exemption from liability of the library, archives office or museum herein provided for shall extend to every officer, agent or employee of such institution in the making and delivery of such reproduction when acting within the scope of his authority of employment. This exemption for the institution itself carries with it a responsibility to see that library employees caution patrons against the misuse of copyright material reproduced photographically.

Under the law of copyright, authors or their agents are assured of "the exclusive right to print, reprint, publish, copy and vend the copyrighted work," all or any part. This means that legally no individual or institution can reproduce by photography or photo-mechanical means, mimeograph or other methods of reproduction a page or any part of a book without the written permission of the owner of the copyright. Society, by law, grants this exclusive right for a term of years in the belief that such exclusive control of creative work is necessary to encourage authorship and scholarship.

While the right of quotation without permission is not provided in law, the courts have recognized the right to a "fair use" of book quotations, the length of a "fair" quotation being dependent upon the type of work quoted from and the "fairness" to the author's interest. Extensive quotation is obviously inimical to the author's interest.

The statutes make no specific provision for a right of a research worker to make copies by hand or by typescript for his research notes, but a student has always been free to "copy" by hand; and mechanical reproductions from copyright material are presumably intended to take the place of hand transcriptions, and to be governed by the same principles governing hand transcription.

In order to guard against any possible infringement of copyright, however, libraries, archives offices and museums should require each applicant for photo-mechanical reproductions of material to assume full responsibility for such copying, and by his signature to a form printed for the purpose assure the institution that the duplicate being made for him is for his personal use only and is to relieve him of the task of transcription. The form should clearly indicate to the applicant that he is obligated under the law not to use the material thus copied from books for any further reproduction without the express permission of the copyright owner.

It would not be fair to the author or publisher to make possible the substitution of the photostats for the purchase of a copy of the book itself either for an individual library or for any permanent collection in a public or research library. Orders for photo-copying which, by reason of their extensiveness or for any other reasons, violate this principle should not be accepted. In case of doubt as to whether the excerpt requested complies with this condition, the safe thing to do is to defer action until the owner of the copyright has approved the reproduction.

Out-of-print books should likewise be reproduced only with permission, even if this reproduction is solely for the use of the institution making it and not for sale.

(*signed*) ROBERT C. BINKLEY, *Chairman,*
      Joint Committee on Materials for Research
W. W. NORTON, *President,*
      National Association of Book Publishers

Cases Cited

Index